THE
STALLION

THE
STALLION

A BREEDING GUIDE FOR OWNERS AND HANDLERS

James P. McCall, Ph.D.

HOWELL BOOK HOUSE
New York

Some of the information given in this book should be acted upon by experienced stallion handlers only. The information is provided for the purpose of education and to give as complete a picture as possible. The reader, even if experienced, should exercise extreme care in all aspects of stallion handling.

Photographs by Lynda McCall unless otherwise credited.
Illustrations by Steve Eaves.

Macmillan General Reference
A Simon & Schuster Macmillan Company
1633 Broadway
New York, NY 10019-6785

MACMILLAN is a registered trademark of Macmillan, Inc.

Library of Congress Cataloging-in-Publication Data

McCall, James P.
     The stallion : a breeding guide for owners and
handlers / James P. McCall.
          p.   cm.
     Includes bibliographical references.
     ISBN: 0-87605-987-6
     1. Horses--Breeding.   2. Stallions.   I. Title.
SF291.M38 1995   94-24194
636.1'082--dc20   CIP

Manufactured in the United States of America
10 9 8 7 6 5 4 3

CONTENTS

FOREWORD

If your goal is to manage stallions, this book will help you benefit from the experience and insight of one of the most competent stallion managers and physiologists I know.

The horse business is the business of horsemen, and in this book Jim McCall gives clear definitions and examples of both sound horsemanship and sound business principles as they apply to stallion management.

I was a student of Jim's several years ago and credit him with showing me more horsemanship and horse management principles than any other single individual that I have known. I recall one lecture he was giving to a large Introductory Horse Management class when his lecture plans began going somewhat off course. The students began asking a lot of very basic questions about why horses do this and that and how do you get horses to do this and that, when Jim stopped all the questions and replied that if people really wanted to learn about horses all they really needed to do was *watch* and *observe* them. Watch them eat, drink, and interact socially. Watch their courtship behavior, and try to see what the horse was telling you or showing you, and from those observations, learn to draw your own conclusions with the horse acting as the teacher and you acting as the student.

That's vintage Jim McCall. In this book, Jim shares several of his experiences and what he learned from them. His experiences with difficult stallions sets the stage for you, whether an experienced or novice horseman, to initiate a young horse into a breeding program or to redirect an older horse toward a safer and more workable situation.

Jim McCall blends experience with reflection in a way that many other writers do not. He captures the kindness and courtesy of others through stories told in a detailed and mannerly way and which are not only enjoyable but also educational.

The economics of stallion management, which Jim McCall presents, is more than "cowboy economics." Current stallion owners as well as others considering acquiring stallions, whether these are stallion shares or percentages, would be well advised to pay close attention to his formulas and cost/return ratios. They are correct. These formulas apply to the Thoroughbred market at present, and I'm sure are appropriate for the Quarter Horse market as well. Furthermore, I strongly suggest you listen to Jim when he includes the tangential thought process the ever optimistic and hopeful potential stallion purchaser may be going through prior to purchasing a stallion. Jim's horse sense bears keeping in mind.

Jim McCall is the product of a lifetime spent working with and learning from horses. In addition he is a natural teacher. He has an innate ability to stimulate thought from facts that have been presented in such a way that students are able to formulate his conclusion. *The Stallion: A Breeding Guide for Owners and Handlers* follows the same thoughtful and logical mode as his classroom teaching. The logic which Jim uses in the selection and management of stallions is well documented and supported by his scientific and physiological observations on semen collection. However, I will add that I believe Jim took semen evaluation procedures of Belgian Draft Horses to a new level when he started working on the amount of recovery time required between ejaculates. He aptly showed that he was certainly flying by more than the seat of his pants.

Jim McCall's observations and experiences became a great part of everyone he touched. The publication of this book is a testament to him and his ability to make horses work for him. Oftentimes I have met with situations and found myself asking, "What would Jim McCall have done?" Jim's lessons give strength, encouragement, and conviction.

Stephen E. Johnson

Margaux Stud
Midway, Kentucky

PART I

BREEDING PRACTICES

CHAPTER ONE

DISCIPLINING THE STALLION

As dawn breaks the darkness of the eastern sky, a couple of cowboys stand warming their saddle blankets next to the campfire. A cold-back bronc raises his head up high as a single cowboy walks toward him to mount in the cold morning air. A fight is imminent. Horse or human . . . who will dominate?

This is a painting I'm describing by the famous western artist Charles Russell. It is called *When Horses Talk War,* and it epitomizes the struggle of will between horse and man. Yet this scene barely touches on the poignant drama that occurs when stallions talk war.

Stallions are the most aggressive members of the species called *Equus caballus.* They were selected over eons for heart, aggression, and speed—traits that can kill in the breeding shed. Discipline and domination by human masters is not easily accepted, especially during the primordial call to procreate. Perhaps this is why nearly every stallion seems to have a story about how bad he is. While reputations vary, consider the tale that accompanied Executive Officer.

A son of the well-known rogue Native Dancer, Executive Officer bit and pawed his handler unconscious. While the man lay on the floor of the breeding shed, the stallion dropped down and rolled on top of him, crushing the man's pelvis. Then, to add further insult to injury, Executive Officer proclaimed his triumph by urinating on him.

A bad stallion? Yes! But, actually, there are very few really mean stallions. Most stallions are made bad by inexperienced handlers. Other than Executive Officer, I can recall meeting only one or two truly tough stallions in my life, one of which was Fourable Joe, a registered Quarter Horse by King and out of a daughter of King. Joe was a small horse, standing about 14.2 hands, but what he lacked in size he more than made up for in guts. He had been the route of most bad actors and had seen some of the best stallion handlers in the country since he tore the chest muscles of his owner as a yearling. Joe seemed to hate people. If anyone walked by his stall, Joe would bare his teeth and charge the screen. With the steel mesh in his teeth, he would proceed to shake the grid as he leered at any human presence with a look in his eye that projected his innermost thought: "If only this steel were your flesh."

I'll never forget the first time I had to bring him to the breeding shed. To say I was shaking in my boots would indeed be an understatement! I told my assistant to close the door behind me and not to open it until I had the horse haltered and ready to lead to the breeding shed.

As I approached this magnificent athlete, all the stories of the men he had supposedly maimed or killed rushed through my mind. Trying to regain control of my senses, I established eye contact with the notorious stud. It was like peering into the eye of the devil himself. There was no respect for my powers as a human being.

Originally, my plan had been to show no fear and to dominate this awesome individual with my courage and determination, but with one look I knew that this approach was lost to reality. I was a young, inexperienced stallion handler and this horse was a fifteen-year veteran of many wars and conquests. Instantly I needed a new approach if I were to remain uninjured and in control of my life. I lifted the breeding halter with the chain-attached leather shank. My demeanor became more conciliatory. I said, "You must allow me to halter you and take you to the breeding shed, otherwise you'll stay in the stall until they hire someone else." The old warrior looked at me again, this time with calm and understanding as he walked over and stuck his head in the halter. I yelled for them to open the door. Joe and I emerged and sauntered toward the breeding area. His attitude reminded me of the time my older brother had to take me along in order to go out with his girlfriend on a school night. I was just along for the ride.

As time passed, my reputation grew as the stud man who could handle the bad ones. And ol' Joe played his part. When tourists came to see the little horse with the black heart, he would still charge at the people like he was going to rip their faces off. Then I would go in and "fearlessly" lead him out. Of course, when Joe told me it was time to put him back up, I knew better than to argue. Obediently, I would explain that it was dangerous to have the horse out too long because he may become aggressive and attack one of the visitors.

Were these stallions born bad? It's hard to say. Too often, the domination of a stallion is seen as a way of proving one's manhood. Stallions are picked at and tormented until some come to view humans with the same contempt once expressed by conquering armies: "The only good enemy is a dead one."

To get along with these powerful, temperamental horses I have developed a code of coexistence. It begins with rule number one:

NEVER TRUST A STALLION

I know, almost everyone can name a stallion that has been a perfect gentleman his entire life. In fact, I have had one or two myself. "Excellent disposition" is usually one of their advertised claims to fame. I still never completely trust them. When something pulls a stallion's primordial trigger, he tends to forget his humanization. How fast he reverts depends on his training.

An overanxious stallion is being reprimanded for being too aggressive upon approaching the mare.

To train and maintain discipline over a stallion, you have to convince him you are more powerful than he is. There are many forms of restraints available, but when it comes right down to it, a stallion is only going to respect the person who is more dominant than he. That is the law of nature.

A common method of stallion control involves a lot of beating and banging. Inflicting pain can give you control over his body, but a stallion can never be dominated unless you gain his respect. A stallion respects power, aggression, stamina, and cunning in a fight. Every day can be a fight with a stallion to see who rules, which leads to rule number two:

NEVER IGNORE A STALLION

Whenever in the presence of a stallion, be at your sharpest. Even the nicest of stallions is trying to play some kind of mind game with you to gain a little dominance. A stallion must have a very clear idea of exactly what he may or may not do.

Draw the line and when he steps over, knock the dickens out of him. Notice that I said *when* he steps over, not *if* he steps over. He will, especially if you don't notice him pushing on the line. Take, for example, the most prevalent bad habit of stallions: biting. A well-trained stallion has definitely had the fire knocked out of him for nipping. He knows it is unacceptable behavior. But do you let him rub his muzzle on you? Was it your idea or his? Unless the stallion clearly understands the answer to this question, he perceives this behavior as a little way to cheat on your rules—to gain a little dominance.

This is the game stallions play. The variations on the basic theme are endless. There is nothing unusual or aberrant about it. A stallion constantly seeks to control his environment—to take charge of a herd. As long as the horse is consistently dominated, the stallion can live in harmony with his human companions. The trouble begins when the stallion controls the situation part of the time or, heaven forbid, all of the time. This brings us to the third rule, which should be followed to the letter:

WHEN IT IS TIME TO DISCIPLINE A STALLION, KNOCK THE HELL OUT OF HIM. NEVER, NEVER PICK AT THE HORSE.

Black is black. Wrong is wrong. Once a stallion understands what he can and can't do, discipline and, if necessary, appropriate punishment is an integral part of keeping the horse under control. I am not advocating beating stallions. The key to gaining their respect lies in the statement made earlier: A stallion must understand what he can and can't do. Like all horses, stallions are not born knowing human etiquette. They must be taught. Fortunately, the herd gives us a hand. A horse learns to expect retaliation if he threatens another animal. Biting, kicking, and other aggressive gestures may bring on war in the pasture. In my barn, the same laws apply. I demand respect from a stallion and in turn I never

Awaiting a signal from the breeding crew that is preparing the mare, this stallion is poised and under control.

take him for granted. I treat him as if I am the herd leader and he is a younger male; I allow him to breed with mares in my band as long as he obeys my rules.

This attitude makes sense to stallions. In the wild, young stallions are subordinate to the breeding males and have a hierarchy within their own male groups. Assuming this position with young two- and three-year-old stallions entering the breeding shed for the first time is relatively easy, as they usually are not willing to commit to an all-out war for domination of the herd.

As you might expect, older and more experienced breeding stallions that have not been taught manners are going to put up a good fight to maintain control in the breeding shed. Their evolutionary history dictates that they should have to fight for breeding rights. Still, most stallions know when the battle is lost and will accept domination by humans who allow them to cover mares.

However, there are a few stallions like Executive Officer who think otherwise. As soon as Executive Officer stepped into sight of a mare his eyes would roll back in his head.

A stallion can learn to accept domination by humans during breeding rituals.

Then, at a full-blown run, he would charge the mare, mount, and begin to savage her as he bred her. Strangely enough, as long as no one tried to interfere with his breeding routine, Executive Officer didn't mind human attendants for the event. God, however, had better help the poor soul who tried to dictate any conditions for breeding. Get in the way, and Executive Officer would try to bite, paw, kick, or otherwise maul you. Pick a fight with Executive Officer and it was a stallion fight to the death. Everything became fair. You had better not blink, for he would nail you. Turn your head and you were dead.

Can a horse like this be retrained? Yes. Yet be advised that the stakes in this savage game of domination can be very high. The loser, be it horse or human, may die. Is the stallion worth the gamble? Not many are, but I took on Executive Officer with the understanding that I would do my best not to kill him. To gain some leverage over this fourteen-hundred-pound stallion that felt no pain as his eyes disappeared, I used a war bridle made with a steel noseband. The chain from the stud shank was run through a loop at the back of the noseband. So armed, I entered the breeding room with Executive Officer forty feet from a mare prepared for breeding. The screaming, stomping, jumping, and kicking increased with each step. Every time "X O" took his attention off of me, I rattled his cage and backed him up. Attempts to break and run were nipped in the bud by spinning him around. After a minute or two, war was declared. The clamor was continuous. Timing had to be perfect. If I misread the horse for even a tenth of a second, the stallion would roll his eyes back and charge. In this state, nothing short of a buffalo gun could stop him. Staying with him in the charge, I tried to make my presence known by snatching and squalling. Mainly, I was just biding my time till he reached the mare. There I could fight again.

As he began to mount, I moved to his flank to prepare for my attack of his underline. As the stallion exposed his belly and other sensitive parts, I delivered a stout blow from the toe of my boot. This usually brought the "breeding machine" back down to the ground. And it put me in a good position for the next mount. As he raised his front legs off the ground, I tried to take the slack out of the line by hurling my two hundred pounds toward his tail end. The leverage pulled him over and I tried to take as many shots as I could before the rogue regained his upright position. Did Executive Officer tuck his tail after losing each scrimmage? No! He came at me with teeth bared and blood in his eyes.

It took about two hours for the horse to decide that it was no longer to his advantage to "flip out." While fatigue was taking its toll on both of us, I was getting more points than he was and I had no intention of quitting. And he knew it! This is the first step in getting a rogue's respect. You have got to convince him that you are willing to fight him to the death. Things will be your way or they will be no way at all.

Now I had control to within twenty feet of the mare. Still, quitting was not in my game plan. I had to get the stallion to yield enough so that the next day would not be a total repeat of this day's trauma. Three hours into the battle, "X O" seemed to make a last-ditch effort to mount the mare. I yelled to the breeding shed attendant to hand me a twitch. I swung toward his head. The blow distracted him enough to enable me to pull him off the mare.

On the next approach, Executive Officer held together to within five feet of the mare. I let him breed. After he dismounted, I walked him back to within five feet of the mare and let him breed again. And again. This was the end of the first day in the Seven Day War.

A cruel, bloody story? No. This is the raw perception of life for the rogue stallion: dominate or be dominated.

Is it possible we are breeding too much aggression into our stallions? I don't believe so. Most stallions are nice horses that would rather not hurt anyone. For the 1 percent with deserved reputations for being rogues, I believe we are not necessarily looking at a problem of too much aggression but of too little horsemanship. Some of the most successful Thoroughbred breeding stallions in the nineteenth century had reputations for being a bit roguish. The English stallion of the late eighteenth century, St. Simon, founded a dynasty that today belongs to the clan of Princequillo. The family of Fair Play, his sire Hastings, and many of his sons were also known for their tempers. The great energy of Native Dancer was also passed down to many of his sons, who were considered cantankerous to handle. It would be hard to imagine the modern-day Thoroughbred without the influence of these horses because it takes aggression to win races. In fact, it takes aggression to do most of the high-energy competitive events: racing, cutting, jumping, and endurance.

I believe that what we need is not less aggressive stallions but more people capable of understanding the uniqueness of the stallion in the breeding shed. A good stallion handler does not walk in off the street. It takes years of experience to understand the nature of the beast. It takes a certain kind of human capable of projecting the power and domination necessary to control a fourteen-hundred-pound breeding machine responding to his hormonal urge to breed. This book is dedicated to those who respond to this calling.

CHAPTER TWO

HANDLING THE STALLION

Hand breeding stallions is one of the most potentially dangerous endeavors associated with the care and raising of horses. The unpredictable nature of these thousand-pound athletes can create some hair-raising situations. I have seen nice horses turn into vicious stallions loathing the smell of humans during courting rituals; stallions driven to rage by the presentation of an unwanted mate; and stallions who lull their handlers to sleep by their exemplary behavior only to attack at an opportune moment.

Therefore, my first rule in the breeding shed is safety. All procedures should provide maximum safety for the stallion, for the mare, and most importantly, for the handlers. Therefore, 99.9 percent of the time, I insist on placing a double scotch hobble on the mare. Ideally, this will protect the stallion and the stallion handler from receiving debilitating blows. While the goal of the breeding shed personnel is to bring the stallion to a mare that is ready to receive him, you can't always be sure that she will agree with your opinion.

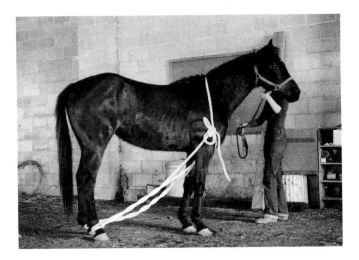

Restrained in double scotch hobbles, a mare cannot deliver a debilitating blow to the stallion.

And unless you have seen it, it is hard to imagine the uproar an uncooperative mare can make even with hobbles on.

My early days at Texas A&M exposed me to a few experiences with fresh-off-the-range maidens—mares that weren't halter broken and had little intention of taking the easy way out when it came to dealing with man. The worst mare I ever had in the breeding shed was a Thoroughbred who came in to be bred to a son of Nasrullah (one of the greatest Thoroughbreds of this century). In spite of the stud's fancy credentials, this mare became terrified as the stallion came into view. She proceeded to throw a walleyed fit, busting out of the hobbles, and driving three people out of the breeding room before clearing a five-foot gate to escape to the backside of a fifty-acre field. Although such instances are rare, it is impossible to predict the behavior of a mare arriving for the first time at your breeding shed. This is why I demand an alert attitude from everyone present.

Still, in spite of the odd mare, the creator of most of the problems in the breeding shed is the stallion himself. Because of this, I am a firm believer in discipline and control during the breeding routine. I expect my stallions to enter the presence of a mare in an orderly manner. Calling, prancing, and dancing is acceptable only as long as the stallion keeps one eye on me and does not take the slack out of the shank.

I believe that hand-breeding routines should closely mimic nature. The stallion should first approach the mare's head. Sniffing, snorting, screaming, and striking the ground are to be expected, but upon command the stallion should back away. The handler should then direct the stallion to approach the mare at her shoulder. More sniffing, licking, and nipping will occur as the stallion works his way to the hindquarters of the mare. During this entire procedure, the stud is expected to back away from the mare upon the direction of the handler and should never be allowed to take advantage of the restrained mare. At the completion of the teasing routine, the stallion is allowed to approach the mare from the rear and a cue to mount is given.

How do you get a stallion to behave like this? Start him young. First-time breeders are easier to train. To make the

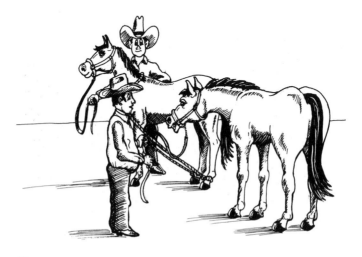

The positions of the mare and her handler and the stallion and his handler are very important for reasons of both safety and control.

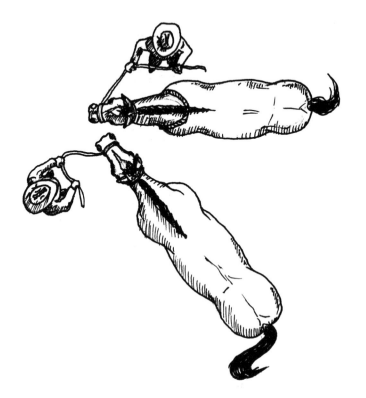

The stallion should be led at an angle toward the mare's head. Upon the exchanging of breath, many stallions strike out, so it is important that both handlers be positioned out of his strike zone.

job easier, it would be nice to let a group of old pasture-bred mares teach the impetuous youngster some breeding manners. Pasture breeding will teach even the most overzealous macho stud not to take mares for granted. There is a natural code of behavior for mating and most mares can get quite angry if they are not treated accordingly. They expect the stallion to declare his intentions and tease her until she hits a breeding stance. Without the proper social amenities, a stallion is likely to be fired upon and rejection will continue until he learns to court the mare according to her wishes.

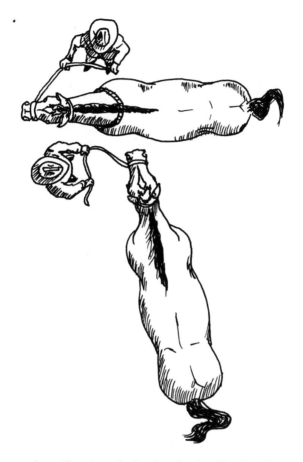

To move the stallion from the head to the shoulder, back him away from the mare and then reapproach.

Unfortunately, most stallion owners cannot afford to take advantage of this method for starting the young breeding horse. With the high cost of quality breeding stock, the inherent risk of permanent damage to the inexperienced stallion is too great. So the lessons designed by nature to be taught by the females of the herd fall solely into the hands of the stallion manager. He is the individual who must provide the rejection or punishment that the mare normally does when a

If all goes smoothly, the stallion is backed off again and led to the flank position. It is at this point that most stallions become even more excited and, if the mare is in good standing heat, she will start to lean toward the horse. The stallion must be kept under control and must not mount the mare prematurely.

rowdy youngster steps out of line. Fortunately, the young breeding stallion's genetic program sets him up for repro-gramming, but it takes force—the kind and amount of force a mare would use. The task is not easy or for the faint-hearted, but done right, it won't take too many sessions to convince the young horse that man is not his enemy trying to steal his mare. Instead, he will quickly figure out that responding to

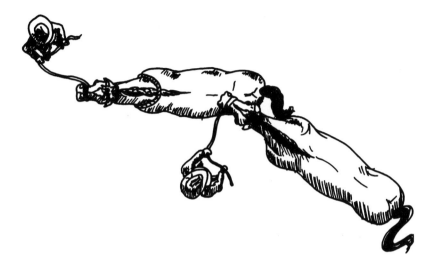

For the third and last time, the stallion is backed away from the mare and returned at a slight angle to her hindquarters. After receiving a cue to mount (such as laying the leadline over her back), the stallion rears and mounts over her hip.

the direction of his handler will enable him to satisfy his biological urge to breed. A camaraderie soon develops as man and horse begin to work as a team.

Before attempting this challenge, be sure to invest in some good equipment, which includes a stout halter and a lead shank with a two-foot piece of chain attached to a snap. While the halter may be either webbed or leather, great consideration should be paid to construction of the lead shank. I have a strong preference for leather because it doesn't burn your hand the way nylon and cotton will when a stallion tries to tear away. However, the chain and the snap are the most important part of this basic equipment. Placement of the chain through the halter and the snap connection of the lead shank to the halter are the primary means of control; the chain and snap need to be stout.

Another decision that needs to be made before the stallion is caught is how the chain should be run through the halter. Some handlers, especially on Thoroughbred farms, use the chain over the nose. I am not particularly fond of this because sometimes it's necessary to jerk the lead line pretty hard to control the horse in a bad situation. A chain over the nose can cut him, leaving scars and dents for everyone to see.

Other handlers like to put the chain through the mouth. When you need to use an arm, you are going to cut the corners of the horse's mouth, again leaving visible scars and damage. Chains are also placed across the gums, right above

Obtaining control by running the chain through the stallion's mouth can also cause damage to the corners of his cheeks.

the top teeth, to achieve a war bridle effect. This is a very sensitive area and if you set a hand to that chain, I'll guarantee the horse will feel it.

I prefer to place the chain under the chin. Pressure generated there will get the job done and cause less permanent scarring. I have used the chain in the other positions, too, and

Placing the chain across the gums right above the teeth can produce a great deal of pain and should only be used when absolutely necessary.

probably will again, but I don't use them on any horse as a matter of practice. I try to use the least amount of restraint necessary to get the job done. Once the horse learns what's expected in the way of manners, there is very little to do except to let nature take its course.

The final preparation step before the training session begins is to arm yourself with the correct attitude. A stallion's behavior will be affected by the way that you feel about his inherent nature. A stallion treated as if he is a man-eating tiger will display more of the characteristics you fear, because your attitude will be expressed in the way that you handle him. Someone afraid of a stallion tends to snatch, bang, and yell at him for every little thing—actions that make him mad and therefore more aggressive and harder to handle.

A more productive attitude for handling stallions is to replace apprehension with a healthy respect for their power along with the confidence that you can deal with any situation that may come up. Combine this with the rules for coexistence and a clear-cut image of the breeding regimen, and the final piece of the puzzle involves the amount of proper handling the stallion receives.

One important aspect of stallion handling is oftentimes overlooked. Because of the high cost of these horses, they get isolated in a solitary, padded-cell situation. Their privileged treatment invites a glass menagerie mentality: Look but don't touch. As a result, these stallions tend to develop more psychological problems, which lead to undesirable behaviors.

The more time a stallion spends on the lead shank, being led to and from his paddock, being groomed, etc., during the entire year, the easier he is going to be during the breeding season.

I remember talking to a stallion manager late one summer night in Oklahoma, who told me, "Right now these stallions are not handling well, but you come back here in April or May, when we are in the middle of breeding season, and they will be handling just like geldings. When we're teasing and breeding, that stallion is on a shank in my hand from the time I finish my second cup of coffee in the morning until well after dark. After a few days with that type of association, we develop quite a rapport!"

Truer words were never spoken, but the idea of having a stallion on the end of your lead shank for eight to ten hours a day would be both a mentally and a physically draining prospect for many. Imagine having to maintain control over a poorly trained stallion who rears, bites, or strikes out. The stress would soon take its toll, so my last suggestion before attempting to train a breeding stallion is to have a plan of action to deal with these potentially dangerous situations.

Biting is probably the most common affliction of stallions. Prone to the use of their mouths by virtue of their sex, stallions use their teeth to fight and their mouths to tease. From the time they are little colts, they nip and nibble at each other and *you* if you will let them. By stopping this early, you will save yourself and others from this cute little trick that *will* turn into a painful, savage habit by the time these colts are ready for the breeding shed.

Acquiring an older stallion that already has this dangerous habit demands an even more cautious approach. At one time, I had a stallion that had a really bad habit of biting, not just me but anything that happened to be around. Coupled with this, he had a rotten attitude about the mares he was to breed. He thought that all mares were supposed to stand perfectly still. If one moved, it irritated him terribly and he would reach over and bite her on the neck until she froze in place. A stallion like this can really hurt a mare, and a leather neck pad should be available to prevent excessive damage. On this particular horse though, I used a noseband made from a piece of wire. Every time he tried to open his mouth, the wire would apply pressure on his muzzle and limit his ability to bite. There are also muzzles that can be attached to the halter, but both of these methods are gimmicks and serve to control the habit—not necessarily to correct it.

It is always better to fix a bad habit, but sometimes the vice is not serious enough to be worth the risk or commitment required to change it. This may be especially true of an older horse who has spent most of his breeding life behaving in a certain manner.

I ran into an eighteen-year-old stallion who tried to bite every time you went to put on the halter. The fellow who had been working with this horse solved the problem by leaving the halter on—a viable solution, except that I have never liked leaving a halter on a horse in a stall. To keep my ego from being damaged, I was determined to come up with a way to halter him without getting bitten. At first I tried punishment and reward, but on a horse who had been practicing this habit for fifteen years, it would have taken a lot of punishment to make an impression. It just was not worth it. I decided to hook my thumb under his jaw as I went to put on

Biting can be difficult to eliminate in the older stallion. For a savage biter, the cage muzzle may be the answer.

the halter. When he tried to reach around, I could push his head up with my hand and foil his attempt to nip me on his first try. The nice thing was that once he had tried, this stallion was satisfied. Haltering could then proceed without further ado.

Pawing is another behavior stallions exhibit. They don't have to learn this, and a little can be tolerated because it is a natural response when approaching a mare. However, there are times when a stallion may get up on his hind legs and take a little punch at you. Usually he will rear up and paw like he is annoyed with you for being there. As I mentioned, some horses just feel that people don't belong in the breeding shed. There are many things to do in this situation, but the one that works the best for me is to sidestep as he rears and give him a good boot in the stomach. This will bring him back down to the ground. Rearing can put a stallion in a vulnerable position, so be ready to turn his threat to your advantage.

Bolting and charging at, around, or past you are all behaviors that show a lack of respect. Once you get the stallion's attention, he must realize that you deserve respect by reason of whatever force is necessary and your attitude toward him. Remember, never take your eyes off the stallion. You have got to be ready at *all times!*

CHAPTER THREE

USE OF THE STALLION AS A TEASER

Every year as I teach horse reproduction to college students majoring in equine science, I ask them to memorize a list of ten signs of sexual receptivity in the mare. Included in that list are the following:

1. Frequent urination
2. Contraction of Labia Major
3. Mucous secretion
4. Seeks company of other horses
5. Labia slightly edematous
6. Body odor of the mare
7. Vaginal vascularity
8. Follicle size
9. Cervical dilation
10. Teaser becomes more excited

Most of the time there is no question about these since they appear to be simple and straightforward. Actually, however, things are seldom as simple as they seem—as the following incident illustrates.

One of my students was a young man who was especially interested in mare management. He worked hard as an undergraduate to develop some proficiency in breeding farm skills and, upon graduation, was hired as a stud manager. In this capacity, one of his responsibilities during breeding season was to determine the sexual receptivity of the mares coming to the court of the farm's stallions. To the amazement of the veterinarian, the new kid on the block could identify mares exhibiting split estrus, double follicles not ready to ovulate, and mares with single follicles ready to rupture before they came through the palpating chute.

This impressive display of intuitive knowledge did not go unnoticed. The following year, my former student was offered a position as general manager on a larger breeding farm. When breeding season rolled around, he again found himself responsible for the teasing charts on the new farm. By April, it was evident from the charts and the vet checks by palpation that the person in charge was not doing an adequate job of determining when to bring mares to the breeding shed. I was asked to try and isolate the difficulty, and so I questioned my former protégé as to what he thought the problem might be. How did he explain the difference in his job performance this year compared to his previous success the year before?

His answer was to the point. "Last year I used the same stallion all the time and teased each mare individually. This year I am working with a new teaser and using a group teasing system."

"While I understand there are some subtle differences in these techniques," I said, "I can't see how this change should have made such a difference in your ability to evaluate the reproductive status of the mares in your care. You were the best rookie I ever saw at observing and evaluating mare behavior."

"That's not exactly true," he responded. "I hardly ever looked at the mare. I just watched that old teaser stallion. After spending so much time with him, we developed a good communication system. By paying close attention to him, I learned to let him tell me the reproductive stage of each mare. If the old horse did something that confused me, it usually meant that the mare had a problem. I just interpreted for the vet what the teaser stallion had told me."

It was apparent that my former student had learned one of the ten signs of sexual receptivity in the mare (the stallion is more excited when the mare is in heat), but he needed to study the other nine. Yet he did point out how useful it could be if we were to develop an astute understanding of the teaser excitation level in the presence of mares to be bred.

As an aid to clarifying the teaser stallion behavior, a scoring system has been developed based on the duration and intensity of the horse's interest. By using this system, it is possible to compare the teaser's behavior with that of the mare's behavior as a more precise indicator of the optimum time for breeding.

TEASER STALLION SCORING SYSTEM

0   The horse is brought into the presence of a mare and finds his advances unwelcome. Within a minute, the teaser is either looking for other distractions or wants to eat grass.

Teaser Stallion Score: 0. The stallion finds his advances unwelcome and will quickly become distracted.

1   When the mare is not totally repulsed by the teaser, the stallion may tease in short bursts but with long pauses in between. These pauses may expand in length until the horse is no longer interested. The horse may get an erection but it will not be maintained.

Teaser Stallion Score: 1. The stallion teases in short bursts with long pauses in between.

2   When a mare shows a mild interest in his advances, the stallion may become more excited and very vocal with bursts of stallion squeals and chuckles that may last for several minutes.

Teaser Stallion Score: 2. The stallion becomes more excited and very vocal, with squeals and chuckles lasting several minutes.

3   When the mare shows strong signs of recep-
     tivity to breeding, the stallion becomes
     aroused and passionate toward her. He will
     continually court the mare with intense
     interest. He may give a flehmen response,
     squeal, and paw the ground, after which he
     will feverishly return to the blowing, nib-
     bling, and talking. The stallion will get and
     maintain an erection.

*Teaser Stallion Score: 3. Aroused, the stallion will continually
court the mare.*

4   The stallion is most aroused just before a
     mare ovulates. He is pumped full of adrena-
     line and exhibits more intense characteristic
     behaviors described in number three above.
     It may be almost impossible to divert his
     attention from the mare. He will continually
     tease her and not want to be led away.

Teaser Stallion Score: 4. Intense interest that is almost impossible to divert.

It is important to note that all stallions are individuals and have their own unique style and level of aggressiveness during courtship. For instance, most two-year-old colts do not have the stamina for teasing or the interest that a three-year-old might. Three-year-olds, however, seem to have nothing on their minds but sex and will become very aroused at the slightest exposure to a mare of any type. They are likely to maintain this interest with very little or no encouragement from the mare. Even with older horses, there is a wide range of libido within two standard deviations of the norm. The stallion manager must be able to read and understand the behavior of his particular teaser stallion if he is going to get the most information available to him during the act of courting mares. Well-developed powers of observation are very necessary tools in an efficient and highly successful breeding operation.

Compare the Teaser Stallion Scoring System with one developed for mares.

SCORING SYSTEM FOR MARES

0   Shows absolutely no signs of interest in the stallion. Rejection is expressed by attempts to leave his presence; kicking, squealing, wringing her tail, etc.

Mare Teasing Score: 0. Shows absolutely no signs of interest in the stallion.

1   Passive resistance. Rejects his advances but not his presence. Appears to enjoy his conversation but not his advances.

Mare Teasing Score: 1. Rejects his advances but not his presence.

2   Likes the attention of the stallion but will not stand to be mounted. May urinate, raise her tail, or have contractions of the vulva (winking); however, her behavior lacks intensity and she does not strike a breeding stance.

Mare Teasing Score: 2. Likes his attention but will not stand to be mounted.

3   Adores the advances of the stallion. Will urinate (throw-off), raise tail, wink, and allow herself to be mounted. Increased intensity of behavior.

Mare Teasing Score: 3. Adores the advances of the stallion.

4   Mare becomes aggressive, seeking out the attention of the stallion. Exhibits all the behaviors of number three above and will back up to the stallion requesting breeding.

Keep in mind that mares slip between these categories of behavioral estrus. To calibrate this movement a plus or a minus sign may be added to the score. For example, a mare either moving into estrus from a "0" or slipping out of estrus from a "1" may be designated as a "1–." She does not completely resent the stallion, but she does not find harmony close to him.

Obviously, there is some subjective judgment to be used in scoring mares. The more familiar one is with the individuality of a mare, the more accurate her heat score will be. For some shy, silent-heat type mares, a "4" may be described as an intense gaze at the stallion and a desire to stand one foot closer to him. On the other hand, there are mares that appear never to slip below a "1" even when they are pregnant. They are always willing to stop and chat politely with any stallion passing by. It is for these odd mares that the behavioral scoring of the teaser stallion becomes an extremely important tool in deciphering exactly what is going on with these intriguing and sometimes confusing creatures known as broodmares.

TEASING SYSTEMS

Several years ago, at a Lexington, Kentucky, farm managers' seminar, a well-respected horseman was asked, "What is the most important ingredient for a successful breeding season?" The response was immediate and direct: "A good teaser stallion!" I hope that after reading the last chapter, you would agree with him.

The management of the teaser stallion is critical to making this essential participant into the most useful tool of the breeding season. This is done by developing a teasing program that will maximize the benefits derived from the use of the teaser. In order to reap these rewards, several problems involving the use of a teaser stallion must first be solved.

INTRODUCING THE TEASER

Obviously, all mares are not going to show the same degree of receptivity to the rather explicit suggestions made by this equine Casanova. Choreographing the introduction in a manner that provides protection to the human handlers as well as to the object of his affection is important to all concerned. As is well known, if one of the mares does protest violently, things can get much more lethal. Mares have been known to show their rejection by biting and pawing, but wheeling, kicking, and squealing are the most likely responses to a misplaced stimulus. An effective teasing system must provide refuge for all parties in any eventuality.

THE COSTS IN LABOR AND FACILITIES OF A PARTICULAR TEASING SYSTEM

Being able to tease large numbers of mares expeditiously saves time and reduces costs as long as the facilities for such a system are not more expensive than the labor savings. When designing systems with such savings in mind, it is important not to loose sight of the delicate nature of the event taking place.

Exposure without distractions and allowing for quality time between the mare and the teaser are the primary elements in any good teasing program. This concept was best expressed in 1972, while I was on a tour of northern Florida Thoroughbred breeding farms. At Tartan Farm, then one of the largest and most respected farms in the area, I asked the farm manager which teasing system he found to be most successful for the mares at his farm. He said that just one teasing system would not be adequate in determining accurately the receptivity of all mares. His conclusion was that to have a high success rate, several approaches must be used. This would increase the chances that a system would be found to titillate the fancy of every mare.

During my travels across the country over the past thirty years, I have seen many of the teasing system options at work. While the point has been made that not every system will fit all mares, it should also be noted that not every teasing system is right for every farm. Individual management situations call for specific solutions to this generalized problem. For example, in Southern California, I visited a farm with a most workable teasing system, but one that would be considered extravagant on a smaller scale operation. This program built its teasing system around the notion that one stallion would not be right for every mare. So they had four.

Mares, housed five to a paddock, were hustled out of their fields by a mounted rider and moved down a twenty-by-sixty-foot chute, where three of the stallions were housed in box-type pens. The first stall was close to the entrance, the second about halfway down, and the third was found at the far exit. Each stallion was separated by a space of approximately thirty feet.

Another exceptional part of this male brothel was that each stallion approached his task of teasing with a different level of aggression. The first stallion was loud, boisterous, and animated. The second's behavior was more normal, and the third whispered and did not express a great deal of excitement when the mares came through the chute. It was expected that this last stallion would stimulate timid mares who were frightened by the first two. If all these failed, a fourth teaser was put to work for those mares who were too intimidated by the mechanics of the prior system. For these supershy fillies, a hermaphroditic male was placed in their paddock. When the feminine stallion was observed grazing particularly close to a mare, it was an indication that she was becoming fertile for breeding.

This sophisticated teasing system was a takeoff on the basic cage system, which I saw first used in Texas in the early 1960s and in which a twelve-by-twelve pipe stanchion, six feet high, houses the teaser. This stall is located in a paddock where mares can be brought for free exposure to the aggressive male. If the paddock is a half acre or more, twenty mares can be exposed to the male in a group, and since no one is having to hold the stallion or the mare, there is a tendency for the manager to leave the horses to court for a longer period of time.

The major disadvantage of this technique is that one or more dominant mares may hog the attention of the stallion, driving the timid mares away so that they may not have access to the horse. This problem can be circumvented by removing those dominant in-heat mares from the pen as soon as they react.

As with most situations, the way the cage system is managed ultimately determines its effectiveness. While managing an operation in Louisiana that bred about 125 mares annually, I designed a cage system that allowed for the teasing of fifty mares an hour and resulted in the detection of more than 90 percent of the mares with breedable follicles.

The system worked like this. Mares were brought in, twenty at a time, from outlying fields and paddocks to a 150-by-250-foot holding area. Adjoining this holding area were two smaller corrals, each 50 by 125 feet. Mares in groups of

The teasing cage.

five were moved into the first corral, where they were able to move up to the cage housing the teaser stallion. If a mare didn't show any signs of estrus, she exited through gate number one, which returned her to the field. If a mare indicated that she was interested in the attentions of the stallion, she was moved through gate number two, into the adjoining second corral. There she would be observed further as she could continue to respond to the teaser. Within two hours, all 125 mares had been moved through the pen system and the mares that needed additional evaluation were now housed in the second corral, located closest to the barn and the palpating stock.

While the cage system is quite popular for exposing multiple mares to a single stallion teaser, the equine breeding program at Colorado State University uses a variation of the traditional chute-teasing technique. Here the mares are run head to tail into a chute, and the stallion handler leads the stallion along the chute checking the receptivity of each mare toward the teaser's advances. This system is a group adaptation of the more traditional teasing chute, where a single

Leading the stallion to each individual mare.

mare is placed in a stanchion and the stallion is brought into her presence.

Other farms turn the tables and instead lead each mare, one by one, to the stallion's stall so that he can nibble and communicate with her through a peephole in the wall. If the mare rejects his overtures or if he becomes too unruly, the door can be closed and relations severed. The method does have the drawback of having to handle each mare that is being teased.

Another method commonly used in Kentucky that reduces not only the need to handle each mare but also eliminates the need for special teasing facilities is to lead a well-mannered teaser into the paddocks and pastures where the mares are turned out. This technique usually demands the services of three people: one person with the records to note the actions of each identified mare, one person to handle the stallion, and the third, a very important person with a buggy whip, to direct traffic around the male.

This teasing procedure requires a well-qualified team of participants, since mares in different stages of their reproductive cycle along with mares with foals tend to have a broad

Teasing down a lane of mare paddocks.

spectrum of emotional attitudes toward this male intrusion. Some mares may want to be close and affectionate, while others may wish to drive this uninvited guest from the premises by charging, wheeling, and kicking. The size of the paddock where this interaction is going to occur will play a major role in determining the intensity of the horse talk. Large fields will give antisocial mares the opportunity to move to another spot where they will not be intimidated. Even so, if the stallion handler and traffic director are not adept at their particular task, they will be in a high-risk situation.

This idea of mares not being forced to respond to the stallion is the principle behind one of the systems used at the University of Maryland. In this situation, the stallion has his stall, paddock, or run adjacent to where the mares are at their leisure. The fence or wall that separates the two sexes must be horse high, bull stout, and hog tight, because both the mares and the stallion are free to run and cavort with just a thin

Handling both the mare and the stallion is the most labor-intensive and costly system for a large operation but may work well for the small breeder.

retaining wall between them. For best results, the time period that this free exposure exists should be limited to a part of the day when the animals can be easily observed, otherwise they may use the cover of darkness to express their amorous intentions and during the light of day seem to be uninterested in propagation.

This highlights the final problem associated with the development of a workable teasing system. No matter how well a system may work for the horses involved, unless the information is observed and translated for the rest of the breeding farm crew, the efforts in design fall by the wayside. Remember, the ultimate objective of all this surrogate matchmaking is to keep the breeding manager apprised of any peaking interest a mare may have in becoming a mother.

A useful tool on the breeding farm is a gelding that will serve as a surrogate teaser. This horse was gelded at eight years of age after being used as a breeding horse for five years. When turned out with a group of mares he will court them at the peak of their estrus cycles.

CHAPTER FIVE

UNUSUAL AND UNACCEPTABLE STALLION BEHAVIOR

Unusual sexual behavior occurs throughout the animal kingdom, including among stallions. Although it is tempting to make value judgments about sexual behavior, aberrant behavior among horses is simply a part of the natural variation that can occur. Since handling stallions in general can be a monumental task, it is important to know of these deviations in order to cope with them on a moment's notice.

In addition, it is hoped that this discussion will provide some peace of mind by explaining that these acts have been witnessed by someone other than yourself. This can be comforting, especially during a crazed moment in the breeding shed.

This was illustrated to me the day I was artificially collecting semen from a huge Clydesdale stallion that I had never seen before. Everything was going quite satisfactorily as the stallion served the artificial vagina (AV). Immediately after ejaculation, however, the horse collapsed and fell back, sitting down on his haunches, much as a dog does. His head was dropped between his front legs, which were rigidly fixed straight out in front of him. He looked like he was asleep. I was amazed and shocked. I had never seen a stallion do this before, and I was afraid he had had a stroke. The stress of the moment must have shown on my face because everyone in the breeding area thought that my look of horror was terribly funny. It appeared that this elderly Clydesdale stallion performed in this manner every time he was bred. Within a few seconds, he opened his eyes, lifted up his head, and rose on all four feet as if he had been startled from his nap by the laughter.

Although that unusual expression of stallion fainting behavior presents no apparent threat to life or limb, it is not always the case. Consider the behavior of a French Thoroughbred stallion that I once stood, who would go rigid, pass out stone-cold, and fall very heavily off the side of the mare. Within seconds, he too would recover, but the fall to the ground was dangerous to the stallion and anyone else in the area. For this particular stallion, the fainting spells would end after a week-long binge of daily breedings, that is, until the beginning of the following year's breeding season.

Another interesting phenomenon that most stallion managers must deal with at one time or another is the rejection of a mare by the breeding stallion. For some reason that he usually does not share with the handler, the horse will refuse to court or breed a particular mare. Although many stallions never show this tendency, a few stallions may reject as many as one in ten of the mares presented to them. No matter how infrequently this behavior occurs, it is extremely irritating to the person in charge of getting all the mares in foal.

Tricky, manipulative matchmaking tactics may be called for. Deception is a good first choice. By altering the appearance and smell of the scorned mare, it might be possible to convince the stallion that he is breeding another. This can be attempted by covering up the mare with a blanket frequently worn by a favored consort. A dab or two of menthol on his nostrils may mask the mare's odor and as a final device, the lights can be turned off by the use of a blindfold. As our deceptive plot thickens, we may let the stallion tease a mare that he finds particularly seductive and then, at the last moment, substitute the ringer. In his frenzied state, we hope our difficult stallion will not notice the switch.

When even the best ploys fail, artificial insemination is a solution if the breed association allows it. Using this technique has the added advantage of less risk of injury to the mare and to the handler should the stallion's rejection turn violent. Once the semen is collected, the stallion has no control over its usage.

Perhaps the most dangerous situation concerning mare rejection arises when a stallion is exclusively pasture bred. Should a pasture stallion reject a mare she will not get in foal, as she will not be bred. In addition, if the mare is left in the pasture with this horse, she may suffer bodily harm. The best solution is to find another horse for this mare. It is highly unlikely that because one stallion rejects the mare, the next one will also. In horse society, one is not tainted by rejection.

Do not confuse the overly aggressive breeding stallion with the stallion who rejects an individual mare. Stallions that address a mare in a crude and rough way lack proper courting etiquette. They have often been spoiled by hand-mating procedures that allowed them an unnatural advantage over the mare. The purpose of restraining the mare is to prevent injury to the stallion. Her movement may be restricted with breeding hobbles, or she might have a front leg tied up, and almost all breeding farms do twitch the mare. If the stallion's deportment is not regulated by the handler, the mare is left practically defenseless. This unnatural situation can lead to stallions that charge into mares so hard that they can knock them down. In a more natural situation, a wily old broodmare would wheel her hindquarters toward the uncouth intruder and present him with a good slap of her powerful heels until his advances became more demure and courtly—no matter how receptive to being bred she might be. Man-made situations allow the creation of this type of unacceptable stallion behavior. It is up to the stallion handler to demand that the stallion make proper advances toward the mare.

Biting is another type of roughness that a stallion may exhibit toward his mares. Biting should not be confused with nipping. Many stallions nip at a mare's flank or hocks during courtship. If the mare is uncomfortable with this, the handler can regulate the stallion as easily as the mare could in the wild. Once mounted, most stallions also will nip the mare on the neck in front of the withers. In normal breeding behavior this can be described as a clamp and hold on type of bite. It appears to serve two purposes. It helps the stallion balance and it immobilizes the mare so that she will stand still.

The problem arises when this behavior goes to the extreme. Stallions that maul mares may actually bite so hard as to remove a chunk out of her neck. A reprimand is not a viable solution for prevention, nor is the problem eliminated in a pasture breeding situation. Protection rather than prevention is the best choice. A muzzle will not allow the stallion to exhibit this behavior. A breeding pad over the mare's neck and withers will also provide her with some protection from the savage bites of her suitor.

A common expression of impending glee for many stallions

is released through maneuvers that greatly resemble the "airs above the ground" of the classical Spanish Riding School. As the stallion is being led toward the mare, he leaps into the air and, while suspended in space, kicks out with both hind legs. This movement is seldom accompanied by any aggressiveness toward either the handler or the mare, but it is certainly important not to have anyone behind the stallion as he demonstrates his joy.

In marked contrast to the "Lippizaner leaper" is the slow breeder. This type of horse is the most frustrating for the stallion handler. His behavior can take several forms. The stallion may be aggressive in courtship but will be reluctant to actually complete the act. Such a fellow is a real tease. Other slow-breeding stallions may, when led into the presence of a receptive mare, completely ignore her. Again, do not confuse this type of stallion behavior with mare rejection. During initial rejection, the stallion may appear to ignore the mare, but if she is forced upon him, he becomes ugly about rejecting her. The "unsexed" individual usually will continue to appear uninterested even when he is forced to have body contact with the mare. In some cases, hormone therapy may help a stallion with such an attitude.

Other times, altering exercise schedules may improve libido as indicated by Dr. James Dinger's study conducted at the University of Connecticut. In this study, eight sexually inexperienced two-year-old stallions were assigned to one of two groups. Group A stallions were exercised at a walk and trot, six days a week. Group B stallions remained in their box stalls. At the end of sixteen weeks, the exercise program for the groups was reversed and maintained for another sixteen weeks. Over the course of the entire study, each stallion was artificially collected every fourteen days. During each phase, the exercised group was found to have lower libido scores than the nonexercised group.

In most cases, pasture breeding the slow breeder does not change his behavior. He is still slow to breed, but at least management is not tied up for several hours waiting for the stallion to perform. Although these stallions are considered abnormal by human standards, it is more likely that they just don't fit into our image of how a stallion is supposed to breed. Since most stallions are aggressive breeders, we expect the horse to get the job done as quickly as possible. Slow breeders are not in a hurry. Watch a "slow breeder" work on a mare in a pasture situation. Many of them are teaching mares how to be bred.

One particular stallion that I stood for many years would court a mare as long as she did not show any form of rejection to his advantages. If she did, he would immediately walk off and ignore her. Within minutes, the coy mare would present herself to him. After an appropriate amount of time this horse would begin his courtship again. This cycle would continue until the receptive mare would stand perfectly still and make the exact responses desired by the stallion. Then and only then would the union be consummated.

Perhaps many of these "slow breeders" find it hard to adapt to the uncreative routines used in hand matings. Many pasture horses have very imaginative mating dances. They prance around their consort, impressing her with their speed, grace, and athletic prowess. It is most difficult under hand-mating management systems to let a stallion perform his own unique mating routines. However, the more sensitive a stallion handler is to the individual need for expression in the breeding stallion, the less likely it will be that his horses will develop any unacceptable behaviors in the breeding shed.

PASTURE BREEDING

Although pasture breeding is not a common practice in today's horse industry, it may have economic value for the small breeder or the owner of a stallion with marginal fertility. As with any management decision, the advantages and disadvantages should be weighed before the system is implemented. The advantages of conserving labor and reducing costs are relatively easy to assess; the disadvantages, on the other hand, must be viewed in light of the expertise necessary to make a pasture breeding system a viable alternative.

RISK TO THE STALLION

One of the first factors that must be considered is the increased risk of injury to the stallion. Pasture breeding does increase the odds that a stallion may be injured by a mare's kick. Fortunately, these injuries are usually minor, since it does not take most stallions long to learn to avoid putting themselves in harm's way. Although the chance is slim that a mare would land a debilitating blow, the possibility does exist and must be appropriately measured.

At highest risk for this kind of injury is the stallion that has not been with a group of horses since he was a yearling. Removed from a herd as a young horse, this stallion was prepped for his racing or show career. Then after the bright lights of the winner's circle, he was retired to a stud farm where he was indoctrinated into a hand-mating routine. His isolation from other horses has been complete. It is going to take longer for this horse to readjust to a herd situation and alter his behavior and give mares the respect they want. But it will happen. Survival instincts and herd instincts, in most cases, have not been erased by five thousand years of domestication.

Besides, there is always the opportunity to help the stallion survive the transition. The preferred way to introduce a stallion with this kind of background into a herd situation is to do so in stages. The first step would be to take the kindest,

Although chances are slim that a mare will inflict a damaging blow, the possibility does exist in an unrestrained situation.

*Turning the uninitiated stallion out with a kind, docile
(in standing heat) broodmare should help reduce the risks.*

most docile broodmare, who is in standing heat, and turn them both into a paddock or small pasture for a few days. When the stallion shows signs of being acclimated to his new-found freedom, his territory is enlarged and more mares are introduced.

ADDING NEW MARES TO THE HERD

For the stallion experiencing his first season out in the pasture, the addition of new mares to his herd is usually not a problem. However, sometimes a seasoned pasture breeding stallion will reject late arrivals once his herd has been established. To minimize this occurrence, it is a good idea to introduce new mares when they are in heat and receptive to the stallion. This does not always ensure acceptance into the herd, but it does improve the odds. As noted, should a stallion absolutely refuse to tolerate a particular mare, it is imperative to remove the mare quickly from the pasture because she is at risk of injury.

Another option for late-arriving mares is to catch the stallion and bring him back into the breeding shed to cover the outside mare. This procedure works well, but sometimes there will also be a rejection of the hand-held mare. When using this technique, it is important to try and cover the outside mare as close to ovulation as possible. Many pasture breeding horses refuse to waste a breeding when they know that the mare is not ready to ovulate.

IMPROVED CONCEPTION RATES

Another professed advantage of pasture breeding is an improved conception rate. In reality, this can be a mixed blessing. I recall several breeding seasons when I was having difficulty in getting four or five mares in foal, using either hand matings or artificial insemination. Unable to find any physiological reason why these mares would not conceive, I turned them out with the pasture stallion. In most cases, they

To improve the odds that a stallion will accept new mares into his herd, turn them out when they are in heat and receptive to the stallion.

were pregnant within three weeks. These mares generally could be characterized as being very nervous and temperamental. I believe that returning them to nature for breeding allowed them to relax and let down, thus enhancing their ability to become pregnant. This solution to solving infertility would not have worked had these mares not been otherwise reproductively sound. Turning a mare out with a stallion will not solve problems associated with infections in the reproductive tract, pooling urine, or a host of other anomalies that would be better treated in a more restricted breeding situation. It is important to check mares to ensure that they are sound reproductively before turning them out with the stallion.

The same is true for the stallion. Stallions with fertility problems are just as likely to have a reduced percentage of conceptions in the pasture as in the breeding shed although marginally fertile stallions seem to have a better chance of causing conceptions when allowed to breed at their own volition. Pro Brandy was one horse whose longevity as a breeding animal was increased by pasture breeding. Pro was a Thoroughbred sire of many hunters and jumpers in Virginia and Maryland in the 1960s. At the age of twenty-six, his breeding career appeared to be over. His semen was still exceptionally fertile, but his advanced age had created a couple of problems. First, Pro developed a problem with balance. As he mounted and tried to seat himself on the mare, he would lose his balance and fall off. This was exacerbated by his second infirmity: The head of his penis would become engorged with blood (balloon) before penetration, making it difficult to enter the mare. After much discussion and many unsuccessful attempts to compensate for his problems, the old stallion was declared sterile due to old age. A Virginia horseman with a band of nine mares purchased the horse, turned him out into the pasture with them, and over the course of the next two years Pro sired fourteen foals before dying.

HERD FORMULATION TIMES

What calendar date is best for establishing a breeding group? It would seem that the optimum time would coincide with the natural breeding season for horses in North America: April, May, and June. If this is true, the next question becomes, How soon before this time can you start? The best answer I have came as a surprise to me. In the past I generally turned bands out the first of March, except on those occasions when I collected the hard breeding mares near the end of the breeding season and turned them with the stallion in a last-ditch effort. In the mid-1970s, however, I decided to try something new. I had two stallions that were standing to outside mares, plus a band of fifteen barren broodmares. I decided to turn seven or eight mares out with each stallion at the end of January and leave them until the end of February, when I would need the studs to start covering the foaling mares and incoming outside mares under normal hand-breeding routines. I did not expect many of the mares to have normal cycles this early, but if they did, I figured I would be that far ahead. The results for the two years that I used this program: an 87 percent conception rate. Thirteen out of the fifteen mares were in foal before the groups were disbanded.

PASTURE BREEDING FOR WET MARES

Dealing with wet mares can require the most skill in a pasture breeding operation. When possible, I like to band the pregnant mares into the breeding herd, including the stallion.

When management and economics permit, these mares are then allowed to foal in the pasture. I believe that there is no more risk from foaling in the pasture with the stallion than there is from foaling in the pasture with a herd of mares minus the stallion.

Actually, the stallion is not normally associated with the risk factor to the offspring of a mare foaling in a herd. The main cause of injury to a foal in this situation is the inability of the dam to protect her new progeny. A foal without a protecting dam will usually be ostracized from the herd. He will be attacked by the other members. It seems logical to assume that nature dictated this behavior so that the liability, a foal without a mother who would attract predators to the group, can be eliminated. To minimize a loss created by such a situation, a good manager should keep close contact with the mares about to give birth in order to ensure that all goes well. Usually, a mare will isolate herself from the band a few hours before she foals. After the foal is up, nursing, and running around, she will return to the herd.

In an attempt to bypass these kinds of problems, some farms wait to turn the mares in with the stallion after they have foaled and are exhibiting postpartum estrus. This seems to work well for them, but introducing a new mare into the herd usually causes bickering. It is this socialization period that puts the foal at risk for getting a swat and possible injury from being in the wrong place at the wrong time.

One way to achieve the advantages of both these practices and reduce the disadvantages is to leave the pregnant mare with the herd during the day and put her in the foaling stall at night. Then, one morning when this mare returns to the herd accompanied by a youngster at her side, introductions are not necessary. She resumes her position in the herd and harmony is still intact. This system closely mimics nature's foaling routine.

SPACE FOR PASTURE BREEDING

The space needed for a pasture breeding operation depends on how many mares you wish to run with a stallion. The minimum of one mare and one stallion requires no more space than a large paddock. The maximum of twenty-five to thirty mares requires a large area, especially if you want to run wet mares in the group. In fact, small traps of less than ten acres are not advisable for herds with foaling mares and/or wet mares. In these small areas, it is difficult for a mare to isolate herself from the group in order to foal. As has been previously discussed, this may jeopardize her newborn foal. Although the actual size to run a maximum breeding herd is dependent on the type of country where you live, a reasonable minimum for this type of operation would be about forty acres, with optimal conditions being between eighty and one hundred acres.

Pasture breeding requires no more space than a large paddock.

NUMBER OF MARES BRED

You can see that the biggest disadvantage of pasture breeding is not necessarily having the space in which to set up the breeding herd, but the fact that there are limits on the number of mares that can be bred to a stallion over the course of the breeding season. In a hand-breeding operation, we know that sixty to eighty mares can be bred to a single stallion. Using artificial insemination we are limited more by the facilities and management than by stallion power, but one hundred to two hundred mares per stallion is not uncommon. With pasture breeding a good ratio would be twenty mares per stallion (plus or minus five mares). We can have the stallion breed more mares, but larger herds lose their integrity and break apart into smaller groups, making the stallion's job more difficult and increasing the odds that a mare will be skipped in any given cycle. Having a stallion that readily accepts new additions to the herd could allow management to take out mares already pregnant and introduce new nonpregnant mares, thus maintaining an equilibrium in herd size that is in the fifteen- to twenty-five-mare range. In this way it is possible to breed forty or more mares in a 120-day breeding season. But if maximizing a stallion's reproductive potential is the primary concern, pasture breeding is not the technique of choice. However, a small breeder who has limited time and expertise and a pasture with good fencing may do well to consider the time-tested breeding system known as pasture breeding.

CHAPTER SEVEN

COLLECTING STALLION SEMEN

One of the best-kept secrets of higher education is that graduate students are given all of the dirty jobs. In the 1960s, the application of this principle found me manning a booth for Texas A&M at a vocational agriculture day held for high school students. Each booth in the exhibition hall represented a specific type of agriculture and was supposedly designed to pique the interest of students who might consider attending college after graduation. Even then, at my tender age and with my lack of experience, I suspected that many of those in attendance were much more interested in missing a day of school than investigating career choices.

As a recruiter for the animal agriculture section, I attempted to lure prospective undergraduates into our tent show. I did so by setting up a couple of microscopes focused on fresh live stallion semen swimming around in the field of view like a mass of miniature tadpoles. Behind me I placed a sign that read, ONE STALLION IS CAPABLE OF BREEDING FIFTY MARES A DAY. I hoped this would hook students into asking what was under the microscopic viewfinder and how that was related to the phenomenal claim of a stallion's sexual promiscuity. My programmed response was to include a discussion of artificial insemination as a modern technique in animal science.

Everything was going as planned until a pair of female students walked up. After a few giggles, they asked an unexpected question: "How did you get what's under the microscope from the stallion?"

At that moment, I developed an instant case of lockjaw accompanied by a mental meltdown. However, the answer to their question should not have been an embarrassment. There are essentially two ways to acquire semen from a stallion: One is to recover it from the female reproductive tract after the stallion deposits it there and the second is to have the stallion deposit the ejaculate directly into a container other than the mare. Historically, the first option was the most likely choice and several techniques were developed to recover sperm from the mare's vagina.

SPONGING

Probably the oldest form of semen collection began by placing a sponge (or cotton ball) into the anterior vagina of the mare before breeding. The stallion was allowed to service the mare; upon his dismount the semen was collected by removing and squeezing the saturated sponge. Although detrimental to sperm life, the technique was perfected by early equine herdsmen faced with a familiar dilemma. Then, as now, too often the most superior stallion was owned by a rival who was unwilling to share the seed.

The solution was ingenious. A tribe member with a mare ready to breed and who wanted the genetic material from his rival's stallion would sneak over and spy on his competitor's mares during the night. His objective was to determine which mare in the rival band was getting the attention of the favored stallion that night. Once the mare was spotted, the daring plan was put into action. The herdsman would sneak up to the mare, place a big wad of cotton in her vagina, and then disappear to patiently wait for the stallion to breed the mare. After the mare was serviced, the thief would again slip down into the herd and remove the cotton. As quickly as he could, he would race to his tethered mare to place the cotton saturated with semen into her vagina. Eleven months later, these crude procedures occasionally produced strong healthy foals—and artificial insemination was born.

SPOONING

Much later, a technique known as spooning was developed to obtain stallion semen. A stallion was allowed to naturally breed a mare. Immediately after the dismount, a technician took a spoon into his hand, entered through the vulva, and scooped the backwash of semen that is normally found pooling on the floor of the vagina. The spoonful of retrieved semen was then carefully carried through the cervix and turned upside down in an attempt to increase the odds that the mare would conceive on that cover (reinforcement breeding).

In spite of the technological advances in breeding equipment that have provided speculums to replace the technician's arm and pipettes and syringes to draw the semen from the floor of the vagina and project it through the cervix into the uterus, reinforcement breeding is no longer a standard procedure around the breeding farm. Recent research has shown the procedure to be totally unnecessary in most instances.

There are, however, two specific situations where the updated technique, formerly known as spooning, actually may increase the odds of conception. The first involves mares who have a problem with cervical dilation during estrus. Upon ejaculation, the small cervical opening restricts the ability of semen to pass into the uterus, which produces an abnormally large backwash. Manually carrying the semen through the cervix will increase the fertility of these mares.

The second condition coincides with an anomaly in penis length. A stallion whose penis is too short to get his glans penis against the cervix of the mare may also fall short of projecting semen into the uterus. "Spooning," or breeding reinforcement procedures, may also help increase the fertility of these otherwise impaired males.

To get a higher quality sample for either artificial insemination or fertility examinations, it is best to use a more modern method of semen collection, such as a stallion condom or the artificial vagina (AV).

CONDOMS

Yes, there is such a thing as a condom for a stallion. On certain occasions when the stallion is unable to be collected with an artificial vagina or it is difficult to obtain such an apparatus, use of a condom may provide the best specimen for evaluation. There are problems, however, associated with getting the condom on and off. And during false mounts and failures to maintain an erection, the rubber cover is difficult to keep in place.

Dave Shaeffer, one of my best friends and an associate in the business of horse reproduction, died as a consequence of this predicament. He was attempting to collect an Appaloosa stallion through the use of a condom. The stallion had a false mount: He entered the mare and dismounted without ejaculation. At this point, the rubber device slipped forward on the penis, resembling a half-pulled-off sock. When the stallion went to remount, Dave bent behind the mare to reach up under the rising stallion's belly to pull the "sock" up snug. The mare, a Quarter Horse named Mexico Miss, had kicked out of one hobble during all this mischief. She rejected the goings on and kicked hard, connecting a solid blow to the solar plexus of this Korean War flying ace, killing him instantly by exploding his heart.

There can never be enough said about taking precautions during intervention in equine mating rituals. A mare being used as the supporting vehicle for the stallion during semen collection should be recognized as the greatest threat of harm to the person doing the collection and should always be properly restrained.

AVs AND DUMMIES

The use of the artificial vagina (AV) allows for more safety precautions and is usually the means of choice for collecting a good quality stallion ejaculate for artificial insemination or for fertility evaluation. One of the main advantages of this method of collection is the option of using a phantom, or manufactured dummy, instead of a live mare—thus eliminating *one* element of danger from this procedure. Training a stallion to mount a phantom and serve an artificial vagina may seem like a hefty task, but for most stallions with a normal sex drive (libido), this is not the case.

When phantoms first hit the market, extreme measures were used to hoodwink the supposedly unsuspecting male into mounting and breeding what amounted to a cloth-covered barrel. Urine from an in-heat mare was splashed over

The mares in the stocks are ready to breed. The stallion is being teased before being led to the phantom.

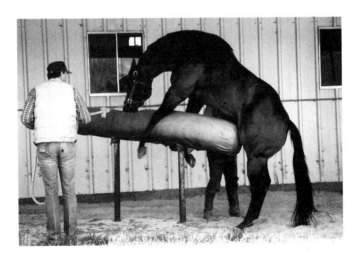

Stallion semen collection using a phantom and an artificial vagina.

the mounting site. The stallion was blindfolded and allowed to tease a mare in standing estrus who was standing next to the dummy. As the stallion was backed off the mare for his final approach to mount, the mare was slipped out of the way and the stallion brought to the back of the phantom. Achieving a mount, his penis was quickly diverted into an AV in the hope that he would stay hooked and provide the collector with the much-desired ejaculate.

While these ploys may still be of use in dealing with the reluctant participant, they are usually not necessary. Most sexually excited stallions will quickly get over their hesitation in mounting the dummy if they find the collection to be a pleasurable experience. But perfecting a model for the artificial vagina and developing the technique necessary to achieve this goal was no easy task.

CHAPTER EIGHT

USE OF THE ARTIFICIAL VAGINA

B efore the beginning of the breeding season it has often been one of my management decisions to collect an ejaculate or two from the breeding stallions. This practice serves a dual purpose: One, it gives a preliminary evaluation of a horse's fertility and two, it provides the opportunity to eliminate the first couple of ejaculates, which are usually of poor quality since the stallion has been out of production for the past six months. This reduction in quality is caused by an increase in the number of abnormal cells. Usually there are a larger number of immature cells that can be identified by the small droplet of cytoplasm found on the sperm's tail, as well as a larger number of fragmented sperm cells contaminating the semen.

I said "it has often been one of my management decisions" because nothing about managing stallions ever falls into the category of "routine." You never know what a simple management decision may lead to—a fact I learned early in my career. One sunny late winter day began as "routine" until I decided it was a good day to collect semen samples from a battery of six stallions. To accomplish this, we needed to produce a jump mare, since a phantom was not available. Around the main barn were several three-acre paddocks, one of which housed an old gray mare who was coming twenty-one that spring. She had lost sight in both eyes but otherwise was healthy and sound and still producing foals. Her best quality (from my point of view that day) was that she was very gentle and never objected to being ridden by a stallion regardless of whether or not she was in estrus. After she was brought up and prepared in the breeding room, I left to get the first stallion, an elderly Quarter Horse named King Rumba, a son of the immortal King and a brother to the great mare Gay Widow.

The artificial vagina was prepared and the two were introduced. King Rumba mounted and served the AV with great enthusiasm. As he ejaculated his body went rigid and he fell stiffly off to the right side of the mare. I had the ejaculate but the stallion lay motionless on the floor. I jumped on his rib cage to try to stimulate his heart and respiration. He gave one final gasp and expired! There was nothing left to do but turn to the old mare and exclaim, "You may be old, gray, and blind but you're still knocking them dead, Sweetheart." An autopsy showed King Rumba's heart had exploded, blowing three large holes in it.

This incident is also a testament to the effectiveness of the artificial vagina, a device that effectively mimics the natural sensations of a stallion servicing a mare. This idea was implemented by the Japanese before the end of World War II. Shortly after their version of an equine artificial vagina reached the market, researchers at the University of Missouri published in 1952 the results of their success with their own design, which became known as the Missouri model. Practically all models available today are based on one of these two prototypes. The AVs are similar because they both have a central hollow rubber tube into which the stallion's penis is inserted. At the end of the tube is a funnel-shaped attachment that empties into a container for collecting the semen. The Colorado model, an adaptation of the Japanese design, wraps an insulation layer around the collecting bottle; a second tube, to be filled with water, surrounds the first layer, all of which is housed inside a rigid hollow pipe. The Missouri model simply places a pliable leather case around the rubber tube.

Both designs have a valve located in the side of the hollow rubber tube, which allows water and air to enter. Manipulation of these elements is necessary to successfully collect stallion semen.

The Missouri model of the equine artificial vagina (AV).

TEMPERATURE

Hot water is the key ingredient that transforms the flaccid tube into a structure that closely resembles the vagina of the mare. The degree of heat and pressure used depends upon the stallion.

During initial attempts to collect a stallion at Missouri, the water temperature attempted to mimic nature. Water heated to 101 degrees Fahrenheit (the temperature of a mare's vagina) was pumped into the device, making it so huge and heavy that it needed to be hung from the ceiling on a cable or propped up on a board stand. More often than not these preparations failed to impress the experimental stallions and they refused to serve the vehicle. It wasn't until the water temperature was increased to between 110 and 115 degrees Fahrenheit that success became more frequent. Researchers learned that the volume of water was less critical than the temperature—a fact that lead to the reduction in size and weight of the artificial vagina.

The lessons learned by these early researchers should not be forgotten. Temperature is still the most critical factor encouraging a stallion to serve an AV. Some horses are so finicky about temperature that they will serve the AV only when the water temperature is within one to two degrees of their preferred heat. I encountered one of these persnickety stallions in Maryland. I began collecting this aged Thoroughbred stallion when questions about his fertility arose. Even through he was in his twenties, he had great libido and would serve the AV with gusto—only to produce a few sperm cells in a small amount of fluids (less than twenty-five cubic centimeters). Our initial reaction was that this old fellow's reproductive system had been compromised by age, but as we later found out, what we originally collected was only the pre-wash fraction of the ejaculate. By raising the water temperature to nearly 140 degrees Fahrenheit (the temperature of hot water from the faucet), the stallion ejaculated 250 cubic centimeters of high-quality semen. This fertile stallion continued siring foals until the ripe old age of twenty-eight, but he would not serve an AV that had a temperature less than 135 degrees Fahrenheit—enough heat to cause most stallions great discomfort.

At the other end of the spectrum, I have never seen a stallion that preferred the AV any cooler than 110 degrees Fahrenheit. Fortunately, many stallions are not as finicky as the old "pro" and will serve the AV with the water jacket temperature between 110 and 120 degrees Fahrenheit.

LUBRICATION

As previously mentioned, the stallion's penis is inserted into the hollow center of the tube, which is surrounded by a water jacket. The sides of this surrogate vagina, being of vulcanized rubber, are irritating to the sensitive glans and prepuce of the penis. To reduce friction, a liberal amount of lubricant should be squirted into the AV and spread down the entire length of the unit. A nonspermicidal jelly such as Johnson & Johnson's KY Jelly should be used.

Once the inner liner is slick enough not to cause irritation, we can deal with another area where stallions seem to show a discriminating preference: tightness of the vagina on the penis, or pressure.

PRESSURE

Pressure can be varied by adding or removing water in the sealed water chamber through the valve-stem plug and by increasing or decreasing the amount of air. Each stallion's individual preference determines how much air and water to use.

Some stallions also resent the fact that there is no end to this hollow tube; a cervix is not present for them to seat against as they thrust. To satisfy those horses, the end of the tube may be pinched closed until ejaculation, giving the stallion the sense that the tube has an end or anterior wall.

To be successful and have the artificial vagina work smoothly, it is imperative to know each stallion's idiosyncrasies for all the sensations that the person doing the collection can manipulate. This is true even for the technique employed to manipulate the stallion into the device. Some stallions are "touch-me-nots," while others are excited by the handling of the penis by a technician.

Each horse has his own personal combination of stimuli that improves the ease and ability to make quick collections of semen. I do not mean to imply that all stallions are in need of complex and detailed strategies, but most collections will be made easier once a horse's triggers have been determined and systematically followed. This consistency also reduces the chance that a horse will develop bad habits due to mishandling (by *his* definition). Bad collection techniques can produce stallions that will refuse to mount a mare if they see someone standing close by, toting an artificial vagina.

This brings us to the topic of logistics: where to stand and how to hold the AV. The collector is normally positioned to the side of the mount—not too far from the action, but far enough off to the side to avoid being struck by the stallion as he leaps up to mount. As the stallion comes forward to seat himself, the technician needs to move in quickly and with his hand direct the penis into the prepared AV. The urgency of this move is somewhat dictated by the composition of the mount. Obviously, a phantom is not going to get pregnant by a slow technique in diverting the penis into the AV; however, it is a different story when the mount is a live, in-estrus mare.

Collecting a stallion from the right side.

Personal choice dictates whether the collector stands off to the right or left side of the mount. My preference has always been to be on the right side, with the stallion handler on the left or opposite side; however, I am probably in the minority today. Still, I prefer this position because it allows the stud handler to pull the stallion away from me and toward him in case the horse has a heart attack or decides he doesn't like my presence and tries to reach over and bite my face. The other advantage for me is that I like to hold the AV in my right arm and manipulate the stallion with my left hand. This is more easily accomplished from the right side. I also press my head into the flank of the stallion once he has entered the AV. Since my vision is obscured during this procedure, this contact between my head and the horse's flank allows me to feel him if he moves. And it allows me to provide a little balance to the mounted horse.

Most of the problems occurring during first-time collections center around lack of patience. It is common for the collector to want to go to the stallion too fast and push the AV onto the penis before the horse is seated. Do not rush the process of entry and do not push the AV toward the stallion until you are behind the mare's hip.

Once the penis has been redirected into the AV, the overzealous technician must be careful not to push a timid breeder away from the back of the mount or be driven backward by a hard-thrusting stallion. The stallion will follow the AV. Being pushed behind the hip of the mare will cause the stallion to back off the mount—a precarious position for the collector and stallion alike. It is up to the holder of the AV to maintain proper position with the AV so that the stallion stays in the natural breeding position.

Note the natural position of the stallion's penis. The AV must be held in such a way that it helps the stallion to maintain a natural breeding position.

Any situation that causes the stallion discomfort or aborts the collection attempt will make the subsequent trials more difficult as the stallion will become unsure or reluctant to try to maintain his position and balance on the mount. Therefore, care should be taken to avert the development of any negative behaviors during breeding maneuvers.

This is particularly important for stallions used in both natural cover and artificial insemination programs. For these horses (as with all horses) we want artificial collection to resemble as closely as possible natural service so that habits do not develop that might interfere with natural breeding. The ultimate goal during artificial collection can be simply stated: Keep the stallion as interested in sex after being collected with an AV as he is after having a natural mating experience. In fact, some stallions that are well managed may actually prefer to be artificially collected.

EXTENDING AND STORING SEMEN

Considering that it takes some powers of persuasion to obtain a semen specimen from a stallion, it is judicious to handle the sample in a manner that does not cause excessive damage to the spermatozoa. Over the last century, many common semen handling techniques unknowingly abused the prized sperm. Spooning, which is recovery of the semen from the floor of the mare's vagina, usually occurred by using a metal spoon washed in soap and water. Fragile equine sperm cells cannot survive contact with cold metal surfaces, nor can they tolerate electrolytes in their environment. Any solution or surface that is a good conductor of electricity (i.e., a metal spoon) is not good for sperm. Soap film is lethal and millions of sperm can be killed simply by coming in contact with the surface of a spoon or container rinsed well in tap water. Only distilled water, free of any electrolytes, should be used to rinse an object coming in contact with the spermatozoa. It is also better to use glass or plastic containers, tubes, and spoons. These minor changes will prevent the assassination of more than half of the sperm cells by equipment with lethal surfaces.

Antiquated techniques are not the only way a sperm can end up in a hostile environment—a fact I learned in a research setting. As I was busily conducting stallion semen experiments, I noticed that another researcher, working in the same lab, seemed very curious about my activities. After several weeks of quiet interest, this gentleman from the Middle East approached me and indicated that he had been doing research on bull semen for some time but had never worked with stallion sperm. He further elaborated that the people in control in his country had a keen interest in good-blooded, swift horses and he thought it would be to his advantage to broaden his knowledge to include the manipulation of the male horse gamete.

I agreed that this sounded like a worthy project for him and he asked me to bring him a semen sample from a particular black racehorse stallion. I tried to brush off the request by saying that I would the next time I collected this horse, but since it was the dead of winter, it would be a while before a sample could be obtained. I hoped that by then the bull semen researcher would forget about the whole thing. Unfortunately, he was much more persistent than I anticipated. Every time he saw me, he asked when I would bring him the sample. Finally, I realized I was going to have to make good on my offer or go back on my word.

So, during a great norther with the chill factor hovering around minus forty, I trooped forth to collect the black stallion. After procuring the sample, I protected it against the cold like it was a newborn infant. As I waited in the lab to be treated for frostbite I handed the sample over to the anxiously waiting researcher. He walked over to his work station and gently poured the raw semen into a goblet full of bull semen extender, promptly killing fifteen billion sperm cells.

(Bull semen is preserved for storage and increased in volume by an extender that contains sodium citrate, an *electrolyte*. While bull semen can withstand this electrolytic buffer, stallion semen cannot.)

The sperm cells of the horse are among the most fragile sperm cells of any species when it comes to survival outside the animal (in-vitro environments). This is especially true when it comes to drastic changes in temperature—an obstacle that is difficult to avoid when collecting semen in the middle of winter. In its natural environment, sperm is kept at approximately ninety-six degrees Fahrenheit—a temperature that is much warmer than the atmospheric temperature during much of the breeding season. Care must be taken to have all contact surfaces warmed to near one hundred degrees Fahrenheit. Keeping the bottles, pipettes, syringes, and extender media in an incubator will reduce the possibility of death due to cold shock, the term used to describe hypothermia in sperm cells. Overexposure causes the male gamete to develop a characteristic appearance: The tail bends backward in an arc toward the head. This causes the live cell to swim in a circular motion (imagine a rowboat being propelled by one oar). This deviant movement prevents the cold-shocked sperm from competing with straightforward moving "streakers" in the race to fertilize the egg. Therefore, a cold-shocked sample decreases the fertilizing capability of an artificial insemination procedure and makes it more difficult to accurately evaluate the fertility of a horse.

Another environmental hazard that needs to be controlled in order to have a viable sample is ultraviolet light. Place a semen sample in a clear container in direct sunlight and within minutes the mass slaughter of sperm cells will begin.

To counter this problem, semen should be collected in containers that have ultraviolet ray filters and then quickly moved into work areas devoid of ultraviolet light.

The next two lethal conditions concern only reproductive physiologists trying to protect the semen sample over a period of time. As the ejaculate sits around waiting to be processed, it becomes its own worst enemy. By-products of cell metabolism change the acidity of the environment. The waste produced by billions of cells becomes life threatening to all the cells in the beaker—a concept we are becoming increasingly aware of in our own habitat.

Also as time lapses, the bacteria naturally found in this nutritive mix of sperm and seminal fluids begin to multiply at an ever-increasing rate, further adding metabolites that reduce the pH (a measure of the acidity or alkalinity of a solution), making the media more acidic. These bacteria are also in competition for the same nutrients as the sperm cells. Together these two factors, left unchecked, will eventually lead to the destruction of the fragile sperm.

For ejaculates to be kept for more than half an hour, counteractive measures to these and other detrimental conditions should be taken. The first line of defense is to cool the temperature of the semen to slow the metabolic rate. To do this and prevent cold shock, the change in temperature should be regulated to drop at a rate of about one degree Fahrenheit per minute until the solution reaches refrigerator temperature (thirty-nine degrees Fahrenheit).

Cold shock can be further reduced by increasing the volume of the sample through the use of a prepared solution (but not bull extender!). The sperm should be placed in the appropriate extender medium before cooling. The correct procedure is to warm the extender and the ejaculate to the same temperature. Mix and then cool the extended semen at the rate given above.

The extender medium can also be designed to help control the deleterious effects of the acid and bacteria. The addition of proteins to the solution will create a mild buffering of the acid. Antibiotics in the extender will help limit the replication of the bacteria.

Remember that time determines which factor is most lethal to sperm cells. Initially, acidity is the biggest problem counteracted by the proteins in the extender. After twenty-four hours, however, significant sperm death can be attributed to the build-up in bacteria populations. Samples that need to be kept this long should be inoculated with antibiotics.

There are other reasons to have more pathogen-free samples. For example, some mares seem to have little resistance to pathogens found in semen. For these mares, it may be desirable to use artificial insemination techniques with semen incubated in an antibiotic extender to retard the growth of these negative bacteria.

Unfortunately, however, our ever-fragile equine sperm cells also find these antibiotics detrimental to their health. So the extender must balance the positive effect of the antibiotics against the bacteria versus the negative effects of the drug on the sperm cells themselves. Concentration, therefore, becomes critical and the concept "if a little is good, a lot is better" will turn life-givers into warriors of death.

Nutrition for the sperm cells and obtaining the proper osmolarity of the liquid are two other considerations in the development of a proper equine extender media. For this reason, glucose water is the standard ingredient in most equine extender recipes.

There are several commercial stallion semen extenders on the market, but if you happen to need one before the next delivery, here is a recipe for an extender that has worked well for me for many years. It is easy to prepare and contains ingredients that can be found in the smallest of towns.

Equine Semen Extender

50 ml of 5% dextrose water
50 ml of nonsweetened evaporated milk (e.g., Pet Milk)
400,000 IU penicillin
100 mg streptomycin

Warm to the same temperature as semen. Mix at a ratio of 1 to 1.

With all that has been said thus far, it seems fitting to reflect upon the old saying, "Ignorance is bliss." Do you suppose those early Arabian tribesmen would have tried to steal semen samples to impregnate their mares if they had the "advantage" of all this modern knowledge? Knowing all that I do after gaining a Ph.D. in equine reproduction, I believe I'd have thought the project preposterous. But, instead, their daring ingenuity produced a new infusion of genetic material in their herd and initiated a bold new concept in animal breeding: artificial insemination.

Over the millennia that concept has evolved to the point where the male gamete of many species of animals, including the horse, can be stored, shipped, and even frozen for future use. Today the stallion seed, designed to live twenty-four to seventy-two hours once deposited in the mare, can be extended and placed in a refrigerator where it is likely to remain capable of fertilizing an ovum for three to seven days.

Even longer storage can be obtained by freezing the sperm at very low temperatures. The major difficulty in this delicate procedure is the formation of ice crystals as the liquid changes into the solid state as the temperature drops to minus 196 degrees Celsius. The addition of an ingredient such as glycerin will prevent crystal formation by causing the liquid to freeze in sheets—a consequence that renders less damage to sperm cells. At best, one should expect that freezing semen will reduce the number of live normal cells by 50 percent, even though it is possible to keep the stallion semen viable for many years.

This number is extremely important in the calculation of the number of mares to be inseminated. The following formula is used to determine the number of mares that can be bred from one fresh ejaculate:

$$\frac{\% \text{ Live Cells} \times \text{Volume} \times \text{Concentration} \times \text{Normal Cells}}{500 \times 10^6} = \text{Number of Mares}$$

For the purpose of this discussion, let's suppose that this ejaculate had 50% live cells, the volume was 100 cc with a concentration of 250×10^6, and 90% of the live cells were normal. Put these values in the above formula

$$\frac{0.5 \times 100 \times 250 \times 10^6 \times 0.9 = 22.5 \text{ mares}}{500 \times 10^6}$$

and we can see that this ejaculate has the potential to breed about 22 mares. Each insemination procedure needs approximately 10 cc; therefore, to breed these 22 mares we need a total of 220 cc. The addition of 120 cc of extender to the 100 cc of raw semen will supply the additional fluids.

(Remember that if the semen was previously frozen, this sample is only capable of inseminating eleven mares—50 percent less than the fresh sample.)

This technology is not at the disposal of all breeders. Each breed association has rules governing allowable breeding procedures. The Jockey Club, which dictates the breeding requirements of registered Thoroughbreds, demands that all conceptions result from a natural cover. Stock horse associations such as the American Quarter Horse Association allow artificial insemination as long as the semen is not stored or transported. Some associations such as the Arabian Horse Registry of America grant permits that allow the use of semen that has been refrigerated, frozen, and/or transported.

SOURCES OF EQUINE BREEDING EQUIPMENT

Jorensen Laboratories, Inc.
1450 North
 Van Buren Avenue
Loveland, CO 80538
(303) 669-2500
(Colorado model AV, con-
doms, semen extenders, etc.)

KBC, International
172-C Fortune Court
Lexington, KY 40509
(800) 928-7777
(breeding rolls, shields and
pads, stallion rings, etc.)

Nasco
901 Janesville Avenue
Fort Atkinson, WI 53538
(800) 558-9595
(Missouri model AV)

PART II

THE REPRODUCTIVE
SYSTEM OF THE
STALLION

ANATOMY

Perhaps you've noticed a stallion in the saddling paddock prepped to race with his testicles pushed up against his abdomen and adhesive tape tightly wrapped around the sack, or scrotum, so that the testes would stay in that elevated position. The rationale behind this preparation might pacify the woman who told me she gelded all her male racehorses because she believed that they ran faster if their medium pendulous testes did not interfere with leg movement.

In actuality, nature has designed for most male mammals an anatomy that protects the testicles and allows for athletic movement at the same time. Under the skin on the ventral two-thirds of the scrotum is the dartos, which is one of two muscles that raise and lower the testes. The other muscle, the cremaster, attaches from the abdominal wall to the thin membrane lining the scrotum. Contracting together, these muscles raise the testes and upon relaxation, lower the testes.

When adrenaline, the fight-or-flight hormone, is released into the bloodstream, these muscles react by contracting and drawing the testicles up against the abdomen (except in the case of some rodents, whose testicles are drawn back up through the inguinal ring, into the body cavity). Horses are not able to draw their testicles back into the body because the inguinal canal normally closes soon after birth. The timetable for the descent of the testes out of the body cavity and into the scrotum is genetically preprogrammed, as is the closure of the inguinal ring, which occurs as layers of the abdominal wall muscles form over the opening.

Of course, sometimes things can go wrong. Occasionally the ring closes before the testes (plural) or testis (singular) descend, creating a cryptorchid, or ridgeling. Should a testis get trapped in the canal as the channel is tightening shut, the resultant strangulation makes rapid movement somewhat painful. This situation creates a slow horse known as a high flanker, or "also ran."

Anatomically, as the testis drops into the sack, it is surrounded by a thin clear membrane of fibrous peritoneum called the tunica vaginalis propria. The same type of material

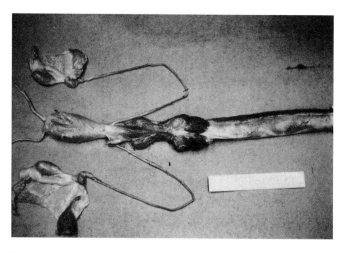

The reproductive tract of the stallion.

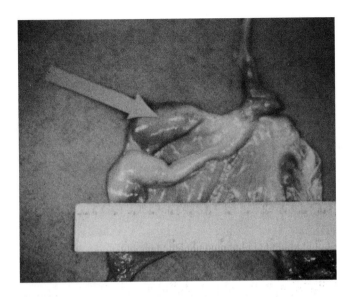

The testis.

also lines the inside of the scrotum and is known as the tunica vaginalis communis. Between these two layers is a cavity and between each testicle is a septum, which separates the testes and divides the scrotum.

Under the tunica vaginalis propria, giving structure and support to the testicle is a thicker, more fibrous layer of white tissue called the tunica albuginea. Many blood vessels can be seen traversing this scar-type tissue. It is this tough tissue that may be ruptured by a severe blow. When this happens, there is an eruption of a pinkish-gray material, the seminiferous tubules, the sperm-producing part of the testis. This eruption will cause a local inflammatory response that will be expressed by severe swelling, or edema, in the testis.

If damage occurs to a large area of the testis, most likely the testicle will atrophy. While this is extremely painful to the horse, his overall future fertility is probably not at risk. In most cases, the uninjured testis will compensate and there will be little, if any, apparent change in sperm production. Nor will there be any reduction in the male sex hormone testosterone, which is produced by the interstitial cells (Cells of Leydig), which are located amid the seminiferous tubules.

Collecting the sperm cells from the tubules where they are produced are the straight hollow tubes, the rete testis. They carry the spermatozoa through the radiating center cord, the

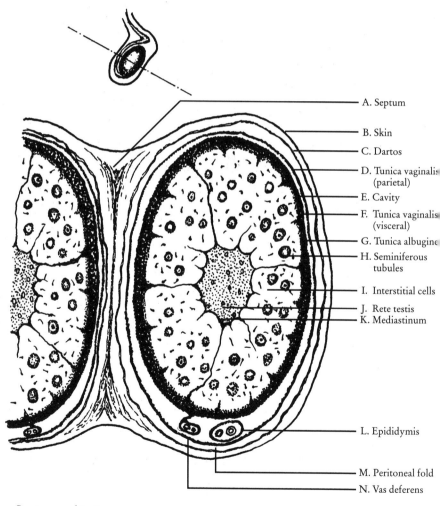

A. Septum

B. Skin

C. Dartos

D. Tunica vaginalis (parietal)

E. Cavity

F. Tunica vaginalis (visceral)

G. Tunica albuginea

H. Seminiferous tubules

I. Interstitial cells

J. Rete testis
K. Mediastinum

L. Epididymis

M. Peritoneal fold

N. Vas deferens

Scrotum and testis.

mediastinum, which empties into twelve ducts at the top of the testis, called the vas efferens. The vas efferens deposits the sperm into the epididymis, which is divided into head, body, and tail. The head, where the vas efferens joins the epididymis, is very closely attached to the testis, but the body and tail are not; in fact, they are often detached from the testis. Of all the common farm animals, this epididymal detachment is unique to the stallion. Should a colt be castrated by a surgeon who is not familiar with this anatomical exception, it is possible that part of the body and tail of the epididymis might be left attached to the horse. It is believed that these tissues can have testosterone-producing Cells of Leydig. Even though the horse can no longer sire offspring, the male hormone produced by the remaining epididymis is enough for sexual desire to remain intact.

Today this hypothesis is not as common an excuse for sexual activity in the castrated male, since we know male behavior can and does exist in geldings that have been properly castrated. Some of this behavior is now attributed to the production of male steroids by the adrenal glands.

The function of the epididymis is the storage, maturation, concentration, and transport of sperm cells as they move through the single convoluted duct. Most of the convolution occurs in the tail of the gland, which forms a knob on the bottom of the testis. This knot is visible when the scrotum is viewed from behind. Upon detection by the uneducated eye, this knob may cause concern. Several stallion owners have asked me to examine a protrusion on their horse's testis, fearing the lump was tumorous. While it is possible for growths to occur in this area, most cases are simply the full firm tail of the epididymis.

Upon leaving the tail of the epididymis, the duct becomes the vas deferens, which transports the sperm upon ejaculation into the body proper. The vas deferens is the

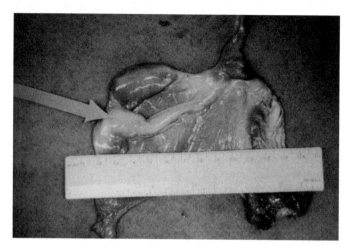

Head of the epididymis.

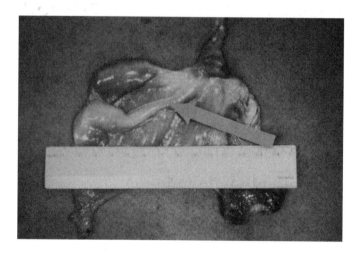

Body of the epididymis.

sperm-carrying portion of the spermatic cord. The spermatic cord is the attachment and support for the external genitalia, i.e., the testis.

This brings us to a second reason why the testes are raised and lowered: thermoregulation. The testis must be kept five

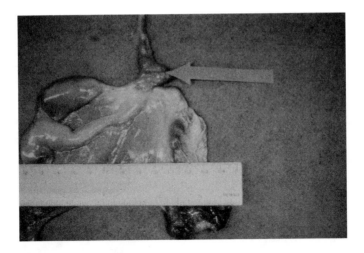

Tail of the epididymis.

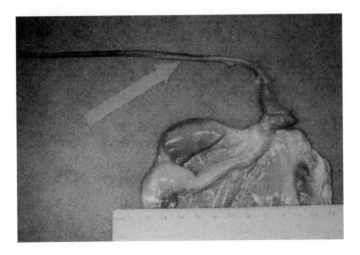

Vas deferens.

degrees Fahrenheit below the body temperature in order to remain functional in the production of sperm. In the case of the colt whose testis did not descend into the scrotum before the inguinal ring closed, that testicle, while capable of producing male hormones, will not produce sperm cells housed inside the abdomen. The environment is too hot.

As a part of my graduate studies, the group studying reproduction strapped the testes of mature rams against their bodies and monitored sperm production. Within a few weeks, a severe decline was apparent, conveying the importance of temperature control to sperm production. The muscles that raise and lower the testes under the influence of adrenaline also respond to temperature. When it is cold outside, these muscles contract to raise the testes closer to the warm body. The reverse occurs when it is hot.

Nature always seems to come up with an efficient solution to an anatomical problem. Yet few can be compared to the

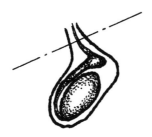

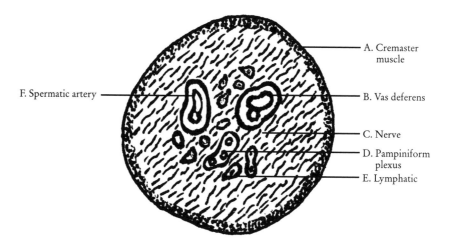

F. Spermatic artery

A. Cremaster muscle

B. Vas deferens

C. Nerve

D. Pampiniform plexus

E. Lymphatic

Spermatic cord.

creative adaptation referred to as the pampiniform plexes—a blood-cooling system for the heat-sensitive testes. Nutrients are carried into the testes by the spermatic artery, which delivers body-temperature blood in such large quantities that it is difficult for the structure to maintain the five-degree temperature differential. The pampiniform plexes is a coil of veins carrying blood cooled by its exposure in the scrotum. This coil is wrapped around the spermatic artery.

The spermatic cord consists of the cremaster muscle, vas deferens, nerves, lymph vessels, blood vessels, and the pampiniform plexus—a rather complex system that is interrupted by wrapping adhesive tape around the bottom of the scrotum. As the spermatic cord enters the abdomen, visual contact is lost and this tube weaves its way through three layers of abdominal muscle, upward toward the pelvic rim. As the vas deferens approaches the rim, its walls thicken with secretory glands and muscles in an area called the ampulla. In the stallion, each of the two ampullae act as a pump during coitus to move the spermatozoa out of the tail of the epididymis, through the duct system, to ejaculation. Passage is improved by the addition of some fluids from the secretory glands in the ampulla. This process enhances insemination by ejaculating the semen with enough force to send it through the mare's cervix, thereby aiding its transport to the target area of fertilization.

Leaving the ampulla on this incredible journey, sperm from each testis are deposited into the colliculus seminalis, an area designed as a mixing bowl, where the sperm mixes with additional fluids. By far, most of these fluids arrive from two pear-shaped lobular glands, the vesicular glands (also known as the seminal vesicles). These glands produce the accessory fluid that makes up the volume of the stallion's semen. Variations in the consistency of these accessory fluids may play a role in the ultimate fertility of a horse.

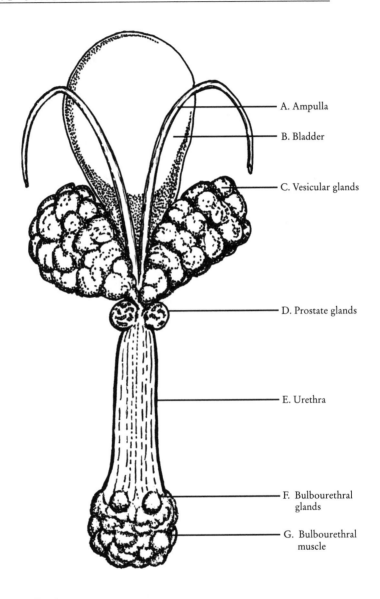

A. Ampulla

B. Bladder

C. Vesicular glands

D. Prostate glands

E. Urethra

F. Bulbourethral glands

G. Bulbourethral muscle

Accessory glands.

Located within the seminal vesicles are the coagulating glands, which provide stallion semen with a unique property not found in most large animals. These glands are responsible

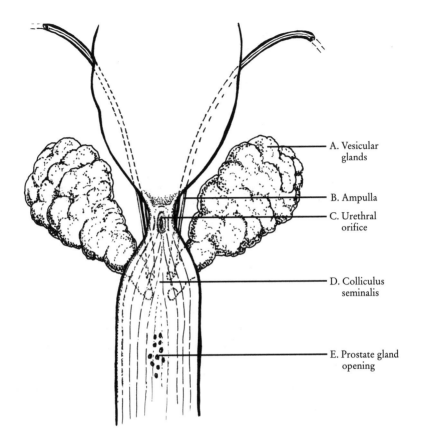

A. Vesicular glands

B. Ampulla

C. Urethral orifice

D. Colliculus seminalis

E. Prostate gland opening

Anterior urethra.

for producing the last fraction of the ejaculate, the gelatinous material referred to as the "plug." The plug is often seen as a stringy viscous substance, expelled as the stallion dismounts from the mare. There are also individual variations in the chemical makeup of this substance and in the amount produced. Nature designed the gelatinous portion to be cohesive and to maintain itself as a separate fraction. In some horses, however, the plug lacks cohesiveness and becomes soluble in the total ejaculate, causing a negative effect on sperm livability.

Despite these individual differences, the amount of plug can be expected to decrease with increased breeding activity and its production to increase in late spring and early summer. Naturally these two factors tend to cancel out one another.

A third gland contributing fluids into this staging area is the prostate. In the horse, there are three rather distinct lobes that function to produce pre-coital fluids that cleanse and lubricate the urethra tube, as they add to the pre-wash fraction of the ejaculate.

The colliculus seminalis is an eminence located on the anterior (front) dorsal (top) wall of the urethra, a tubular structure beginning in the pelvic area and continuing the length of the penis to the exterior. The first six or eight inches of this tube is thick walled with the opening of the bladder, the urethra orifice, located in the anterior (front) ventral (lower) portion.

A pair of oval glands, the bulbourethral glands (Cowper's glands), located in the posterior portion of the thick walled section of the urethra, produce a clear, slick secretion that, like the prostate secretion, serves to cleanse and lubricate the urethra before ejaculation. These glands are embedded in the bulbourethral muscle, which together with the ischiocavernosus muscle, form the root of the penis near the posterior rim of the pelvic arch.

The penis of the male horse is erectile in type, as is that of Homo sapiens. Other farm animals, such as the bull, ram, and boar, have a fibro-elastic penis that is, to some extent, rigid at all times and can be withdrawn when not stimulated. The stallion's erectile-type penis swells with blood when the venous return is restricted by the ischiocavernosa muscle, thus inflating the penis. In its relaxed state, it withdraws into the sheath, the hair-covered skin that houses the penis outside the body.

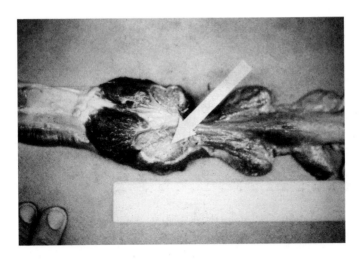

Attachment of the ischiocavernosus muscle.

The penis is divided into three sections. The body, or shaft, is the main portion, comprising most of the length. In cross-section, there is a tough outer covering called the tunica albuginea (the same as the tough outer covering of the testes). Inside this coat of fibrous tissue are two spongy areas that join together on the dorsal side of the penis, the corpus cavernosum. On the ventral area of the cross-section, surrounding the urethra, is another bundle of spongy tissue, the corpus cavernosum urethra. As these tissues engorge with blood they aid in the erection and extension of the penis during copulation.

If a mare were to kick an engorged penis hard enough, it could rupture these bodies of tissue or possibly even the tough outer covering, causing a massive hematoma (leakage of blood). The penis would then become swollen in the area of the rupture and would not be able to fully retract. In severe cases, scarring might even occur that could cause the penis to become deviate (or crooked).

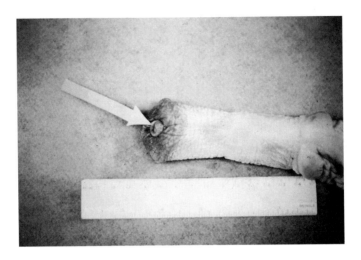

Urethral process.

The terminal portion of the penis is called the glans, an enlarged, blunt, concave body with a short, extended urethral process, on the surface. The glans becomes greatly distended when engorged so that it can seat against the cervical opening in the mare. The urethral process is coupled in such a way with the mare's cervix so that it will allow penetration of the ejaculate through the cervix and into the uterus. However, in some cases, the length of the stallion's penis complicates this process. While little can be done to help a stallion with a short penis seat against the cervix, a horse with an exceptionally long penis can be prevented from bruising a mare's cervix through the use of a breeding roll.

A breeding roll is six to eight inches in diameter with a one- to two-foot-long handhold sticking out of one end. As the stallion penetrates, the stud handler places the roll on top of the stallion's penis between the mare's buttock and the abdomen of the stallion. This will shorten the stallion's penetration by the diameter of the roll, eliminating the discomfort

Breeding rolls.

which invariably causes the mare to try and twist, turn, and walk forward during copulation.

The prepuce is invaginated skin surrounding the glans of the penis. As the penis extends, the folds are unwrapped and provide a covering for the body of the penis. Secretions from the preputal glands act to lubricate and protect this sensitive skin when the animal is sexually active. As activity diminishes, these oily, waxy secretions combine with cornified cellular debris (gunk) and bacteria to form smegma. Small amounts of this substance may be present at all times, but accumulation may occur during nonbreeding periods. Male horses should be checked periodically to be sure that an accumulation of debris has not hardened into a bean, which lodges itself beside the urethral process. A bean will cause irritation and pain if not removed.

For this reason, it is recommended that a gentle cleansing with water occur before the onset of breeding season. In the past, this cleaning involved the use of a mild soap, which was

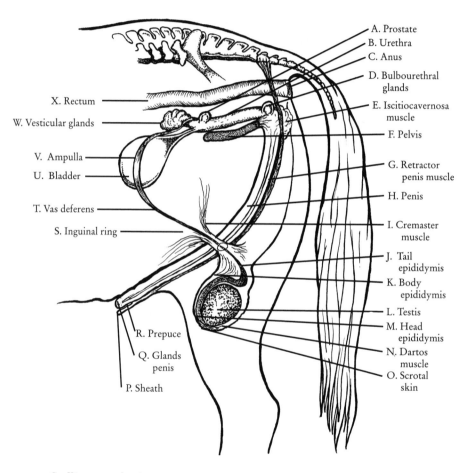

A. Prostate
B. Urethra
C. Anus
D. Bulbourethral glands
E. Iscitiocavernosa muscle
F. Pelvis
G. Retractor penis muscle
H. Penis
I. Cremaster muscle
J. Tail epididymis
K. Body epididymis
L. Testis
M. Head epididymis
N. Dartos muscle
O. Scrotal skin

X. Rectum
W. Vesicular glands
V. Ampulla
U. Bladder
T. Vas deferens
S. Inguinal ring
R. Prepuce
Q. Glands penis
P. Sheath

Stallion reproductive anatomy.

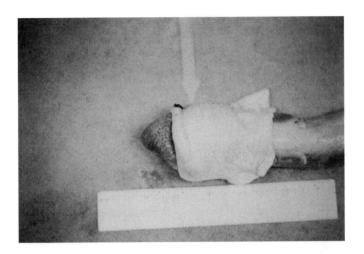

Prepuce.

thought to help reduce infections in the mare and the stallion. More recent research, however, has shown that this procedure has the opposite effect. Removal of the naturally occurring flora of organisms increases the chance of infections by pathogens.

Knowing the parts and the functions of the stallion's anatomy is beneficial when dealing with the male horse, even if you are not evaluating reproductive potential. For example, knowing the necessity of checking for a bean will ensure that all the stallions and geldings in your care will experience less discomfort; owners will not become stressed when they see a newborn colt who, as yet, does not have control of his retractor penis muscle as long as his penis is continuously outside the sheath; and you will be less likely to overlook the development of a knot on the wrong part of the testicle.

CHAPTER ELEVEN

PHYSIOLOGY

As the grip of winter begins to loosen, days of springlike weather lighten the spirits of horses and human beings alike. This is especially true of many old stallions, who seem to sense the impending warmth of spring and become excitable and more interested in the movement of horses (mares, in particular) around the farm. Actually, I believe that all that hype associated with groundhogs and their shadows should be redirected toward stallions and their breeding behavior. After all, stallions are much more aesthetically appealing than groundhogs and their spring-predicting behaviors do not depend upon the sun breaking through the clouds, but rather are partly learned and partly biochemical in nature.

Although it is well known that most mares are seasonal polyestrous (producing eggs throughout several breeding cycles in the spring), it is oftentimes overlooked that stallions are somewhat seasonal breeders as well. The reason for this oversight is that stallions do produce viable sperm year-round, but they also undergo specific changes relative to the season of the year.

The most obvious fluctuation occurs in testicular size, which shrinks and grows in conjunction with the transition from the dormant phase of the year to the active time of late spring and early summer. While this anatomical difference is not as pronounced as in the white-tailed deer, a short-day breeder whose testicles almost double in size with the approach of its breeding season, the hormones in both species are responding to changing photoperiods. In the horse, increasing daylight hours stimulate an increase in the production of the male hormone testosterone, which results in the increased testicular size and activity.

Along with this change, the seminal vesicles also increase the production of their fluids. This is especially evident in the production of the gelatinous plug portion of the ejaculate, which is more copious during peak breeding season and may be practically nonexistent during the winter months.

This cycling of male biological activity starts as soon as the young horse reaches puberty, which should be distinguished from the onset of sexual activity. Somewhere around twelve months of age, colts will begin to mount and ride fillies. Over the next three or four months, they may get an erection and actually penetrate the female. Upon dismount, accessory fluids may be seen dripping from the penis. Still, all this does not necessarily mean that the male has reached puberty.

Puberty is achieved only when the young male is capable of causing a pregnancy; that is to say, puberty is achieved when the male can produce sperm cells capable of fertilizing an egg. In light horse breeds, this usually occurs between twelve and eighteen months. For the larger breeds, puberty may be delayed up to twenty-four months or longer, even though sexual excitement is exhibited in the presence of an in-heat female. To be on the safe side, it is recommended that intact males be removed at approximately twelve months of age from the presence of any females that are not to be bred.

It is impossible to predict the precise day of puberty, as many factors play a role in determining its onset. Genetic programming and nutrition are two of the more important causal factors because they influence the hormonal flux taking place in the maturing body. This activity is largely under the control of the pituitary gland, which is attached to the hypothalamus and located near the base of the brain.

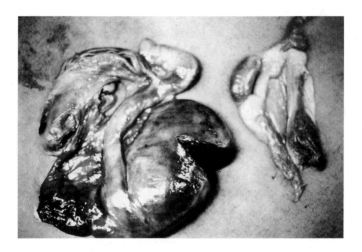

Testicular size is one of the more visual signs of sexual maturity. The testis on the left is of a mature breeding stallion; the one on the right is of an immature one.

The pituitary gland is divided into two portions: the hind portion (the posterior pituitary) and the front portion (the anterior pituitary). The latter controls much of the reproductive nature of the animal through the production of two hormones: follicle stimulating hormone (FSH) and interstitial cell stimulating hormone (ICSH). Both of these hormones are carried through the bloodstream to reach their site of reactivity: the testes. The action of FSH on the testes is to stimulate the production of sperm cells. ISCH causes the

Cells of Leydig (located in the testes) to produce the male hormone testosterone.

Testosterone is responsible for the growth, development, and maintenance of the male reproductive system both internally (the vesicular glands, prostate, and Cowper's glands) and externally (the testes and glans penis). The secondary sex characteristics that give the stallion his macho physical appearance are also under the control of this male hormone. For example, colts develop much more muscle mass than fillies, especially in the neck and shoulders and around the jaws of the face. The crest of the neck, which is primarily connective and adipose (fat) tissue, is more developed in the male equine. It has also been my personal observation that the hair coat of very masculine stallions is different from that of geldings or mares. The hair of these horses has a fineness and luster that prevails even in the winter months when the coats of other members of the species have grown rough and coarse.

Voice is also a secondary sex characteristic that is a verbal indicator of the onset of male hormone production. A similar phenomenon occurs in the males of our species. The higher pitched sounds of the prepubescent male differ considerably from the deeper squalls and guttural chuckles of the breeding male.

Libido, or sex drive, is also related to testosterone. It is thought that the higher the level of circulating male anabolic steroid, the greater the sex drive.

With all of these factors controlled by the same hormone, there should be a relationship between these sex characteristics and the libido of the horse. A very masculine stallion should have well-developed muscles, a crest on his neck, full jowls, a deep voice, good-size testicles, and a seemingly endless desire to cover mares.

CHAPTER TWELVE

FERTILITY

You have found the stallion of your dreams. He has pedigree, conformation, disposition, and a winning record. The price is high but you think you can swing it. The only thing that makes this superior animal so valuable at this point in his career is his ability to stand at stud, so logically you should want his fertility guaranteed.

I am always amazed, however, by the number of breeding stallions that change hands without undergoing a fertility examination. Too often, calls for testing are heard only after problems begin to arise. Take the case of a young man who bought an older stallion that had already sired many offspring. Although the sale's agreement guaranteed the horse's fertility, the horse did not undergo an evaluation of his fertility. The new owner entered the breeding season assuming that he was going to hand breed the stallion, and he booked twenty outside mares to the horse. Only two mares conceived!

Upon completing a semen evaluation, I found that the horse produced only one hundred million to two hundred million live normal cells in his ejaculate. The average horse has about ten billion! This low number suggested that the stallion was the culprit in the reduced conception rate among the mares and that the new owner was eligible to ask for his money back, since the horse's fertility was marginal. As mentioned earlier, in a pasture breeding situation the stallion might be fertile enough to settle a group of ten to twenty mares. The owner decided not to cancel the deal, and late in the season turned the stallion out in the pasture with ten of his own mares. All of them got in foal.

This case also points out that while fertility is usually perceived as an all-or-nothing factor, it is actually a gray area. A horse such as this particular stallion with somewhat questionable fertility can find an owner to whom his bloodlines are important enough that he is willing to pay the price to pasture breed a limited group of mares each year. Still, the comparison between the breeding potential of a stallion and the needs and wants of a new owner should occur before the deal is finalized, because most of the time marginal fertility is not good enough to economically justify today's stallion prices.

As we enter the gray area of what constitutes fertility, there are many evaluations, both qualitative and quantitative, that may be used to give an educated appraisal of a horse's breeding strength. Eight of these evaluations can best be done by collecting the complete ejaculate from the stallion with an artificial vagina. Additional equipment necessary to set up a small semen evaluation lab includes a microscope for viewing the male gamete. The microscope does not have to be powerful or expensive to do a decent fertility exam, but it should have a magnification of at least 10 × 40 with fair resolution. Some sort of graduated container will also be needed to measure volume, which is the first measurement taken on the stallion's output of semen.

When the ejaculate is collected, three fractions are produced: the pre-wash, the sperm-rich fraction, and the gelatinous plug. The sperm-rich fraction is the part with which we are most concerned. The first evaluation is the volume of this part of the ejaculate. A range between twenty-five and 250 cubic centimeters is normal for light horse breeds, with most ejaculates being around one hundred cubic centimeters. (Draft breeds average around four hundred cubic centimeters.)

Within the species there is an inverse relationship between volume and the next parameter: concentration. Concentration is reported as the number of spermatozoa (sperm cells) in one cubic centimeter of semen. In general, as the volume increases, the number of cells found in one cubic centimeter of semen decreases. The normal range is one hundred million to five hundred million sperm cells per cubic centimeter, with the mean being about two hundred million sperm per cubic centimeter. The educated eye of someone practiced in evaluation of sperm concentration can be accurate to within 25×10^6 cells per cubic centimeter by looking at a sample in the field of view of a microscope. For the less experienced, a hemocytometer, which is used to count red blood cells, can be adapted. The hemocytometer is simply a grid-marked slide. This technique, as with the experienced eyeball technique, results in a sample error around 25×10^6.

More precision may be obtained using spectrophotometry. This method provides a reading of the density of a fluid by measuring how much light can pass through the medium. While this is the most accurate measurement of concentration, it does require that the machine be calibrated for sperm cell density; and the price of the machine demands a large number of samples to make the test cost-effective.

Also evaluated under the microscope is the third characteristic of semen analysis: motility. Motility is an estimate of how many cells out of a sample of one hundred show life by movement. Most fertile horses have motility that ranges from forty to seventy per hundred. Said another way, these horses have between 40 and 70 percent motility.

The way these cells move is the fourth characteristic. Subjectively, graded on a scale of one to four (with four being the best), an evaluation is made as to how fast and straight the majority of spermatozoa are swimming. This particular evaluation is considered by some to be one of the most critical in determining breeding potency. The sperm cells of a fertile stallion are expected to have a score of either three or four. What kind of speed does this represent? An anecdote will illustrate.

Shortly after I arrived from Texas A&M to join the faculty at the University of Maryland, a horse became severely injured. He was in such bad shape that he needed constant surveillance until he could regain some strength. I volunteered for the 1 A.M. to 3 A.M. shift. At about 2 A.M. I heard a major commotion coming from the area of the dorms.

Startled, I asked one of the students what could be causing such a disturbance at this time of the morning. The student replied, "Oh, nothing—it's just a streaker!" Apparently, some of the male students had stripped off their clothes and run hell-bent through the dorm area.

It then struck me how to teach students about the rate of forward movement (RFM) of sperm cells. A good RFM is an imitation of a nude student runner covering the shortest distance between two points—moving in a straight line and fast.

The fifth examination is performed on the morphology of the cells to discover how many of the spermatozoa are mis-shapen. Horses have a slightly larger percentage of cells with irregular morphology than other domestic animals.

Normally, up to 25 percent abnormal cells can be tolerated in a stallion without cause for concern, but remember that an ejaculate needs to be evaluated in terms of all of the criteria. I learned this from an eighteen-year-old stallion whose ejaculate was in the normal range for volume, concentration, and rate of forward movement. But he had 30 percent abnormal cells with a motility of only 30 percent. Based on these first five criteria, this horse appeared to have marginal fertility.

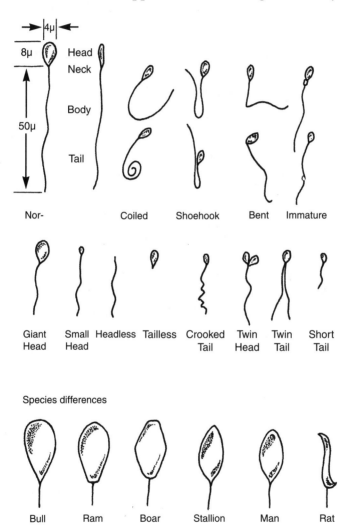

Morphology of Stallion spermatozoa.

Still, thirty mares were booked to him and the stallion greeted breeding season by covering each mare under a hand-mating routine. He settled all thirty with exceptional ease.

The question of what made this horse with limited semen quality such a successful breeder was puzzling. To better understand this stallion's breeding potential, I performed the sixth test, known as the livability test, in which the ejaculate is extended and then stored in a refrigerator. Monitoring then begins to see how long the semen maintains its fertilizing capability. When compared with the normal storage life of other stallions, this particular stallion's semen lived two to three days longer in the in-vitro environment, demonstrating a previously unmeasurable strength.

During this same time, I was standing a second stallion with semen production in the acceptable range on all five characteristics. When this horse's sperm was subjected to the livability trial, the sample was completely dead after twenty-four hours. Since this was about two days less than normal, it was not surprising that this stallion was experiencing some problems in stopping mares from cycling. What was the physical difference in the ejaculates of these two horses that produced such a dramatic difference in their potency? The seventh test provided a clue.

A cryoscope, which records the temperature at which a substance freezes, was used as an indirect measure of semen osmolarity. The weak semen from one stallion froze at a lower temperature than the stronger semen from the other stallion. This indicated a higher osmolarity and a higher electrolyte concentration. Evidently, the seminal fluids of the older horse were more protective of the spermatozoa than the other stallion's. Although most people in the field do not have access to a cryoscope, the information gained may warrant the effort and expense for evaluating high-priced stud power.

Additional laboratory equipment is also required to perform the eighth test. The semen is plated, or cultured, to determine the presence of pathogens in the ejaculate. One pathogen of major concern is the gram negative rod bacteria, pseudomonas. The presence of these organisms can adversely effect the fertility of a stallion, even though the horse seems to be a strong breeder. Pseudomonas is a very difficult pathogen to eliminate from the reproductive tract, which is why this test is one that I would always require of any horse considered for purchase. The prognosis for a horse with pseudomonas is varied. In some cases, there seems to be little change in conception rates; in others, the horse may not be able to father foals. And, of course, all variations exist between the two extremes. Generally, I suspect a horse with a pseudo-monas infection to have a 20 to 40 percent reduction in fertility. This is not the whole story, though.

A stallion testing positive for such pathogens provides a higher risk of introducing these organisms into the reproductive tracts of the mares he breeds. Although bacteria such as staph and pseudomonas are easier to control in the mare than in the stallion (given the techniques and drugs available today), it is still an expensive and tenacious malady to deal with—one that all mare and stallion owners would certainly want to avoid.

The next test of fertility, number nine on this list, is test breeding. A stallion is bred to three to five mares before he enters stud duty. Records are kept to determine the ease with which these mares conceived. The key to a good test breeding program is to be sure that the mares used are capable of conception, so that any problems can be safely attributed to the stallion. A good result for this test would be that all the mares conceived either on the first or second cycle breed.

Test breeding also provides the opportunity to measure a stallion's libido (sex drive). How long does it take the horse to cover a mare? Even though there is little or no correlation between libido and fertility, for most normal stallions being used in a hand-breeding program, libido is the main limiting factor in determining how many mares can be served.

Attempting to quantify sex drive is difficult. Researchers have tried to come up with a subjective scoring system for the expression of male aggression in a breeding circumstance. Others suggest that libido can be measured by the length of time it takes for a stallion to enter the breeding shed, get an erection, and ejaculate. An aggressive breeding horse can accomplish this within two minutes. A slow breeder, however, may take thirty minutes or more. While both horses may be equally fertile, the slow breeder is going to test management's patience, especially if he has a full book of mares and is expected to cover more than one mare a day.

For me, the real quantifying test is the number of mares that a horse will serve in a five-month period. This allows for the evaluation of a stud prospect relative to the needs of different breeders. For example, a horse capable of covering three mares a week would not be a sound investment for an operation hoping to book sixty mares, but would be fine for a small breeder with aspirations of hand-breeding twenty mares or for use in an AI (artificial insemination) operation.

There are still other factors to be aware of when performing a fertility examination, even though they rarely occur. Blood in the semen (hemospermia) will seriously limit the fertility of a stallion. Blood is lethal to spermatozoa and also indicates a problem in the urogenital system. Urine leaking into the semen during ejaculation is definitely cause for concern. Urine is also a lethal substance for sperm cells and almost always causes total sterility. The few cases I have witnessed have appeared in stallions more than twenty years old who developed the problem as a result of the aging process, but cases have been reported in younger stallions.

Clumping of sperm cells is also an indicator of poor fertility. The nucleus of the spermatozoa has an electrically charged surface that causes it to be attracted to particles of an opposite charge. This phenomenon is commonly seen under the microscope when dirt and debris contaminate the sample.

LIBIDO SCORING SYSTEM FOR STALLIONS BEING COLLECTED USING A PHANTOM[1]

Libido Score	Criteria
0	No sexual interest. Unable to collect or breed.
1	Little sexual interest. Slow to achieve erection (more than 3 minutes). Requires mare next to dummy.
2	Moderate sexual interest. Erection within 3 minutes. Requires mare next to dummy.
3	High sexual interest. Erection within 30 seconds. No mare required next to dummy.
4	Intense sexual interest. Immediate erection. No mare required.

[1] *After Dinger and Noiles, University of Connecticut.*

Under the microscope, large numbers of sperm cells will be found attached to the surface of the substances, thus "clumping." When clumping occurs in samples free of contaminants, the horse is likely to have a major reduction in fertility. It has been suggested that the cause of this clumping may be a result of injections of steroids. Whether there is any validity to this idea or not, it is certain that large doses of steroids such as cortisone compounds do reduce fertility. While removal of these compounds from the animal's system should allow for the return of fertility, there are no guarantees. The length of time it takes to reestablish quality sperm production is also unpredictable. Therefore, before purchasing a stallion for breeding, it becomes very important to have a history of his medications.

One of the more famous examples of this situation involved the Thoroughbred racehorse Secretariat, who, during the course of his racing career, received cortisone to reduce inflammation. While this outstanding athlete was in the process of seizing the Triple Crown (Kentucky Derby, Preakness Stakes, Belmont Stakes), he was syndicated for a large sum of money for stud duty at the termination of his career. The syndication agreement required that the horse must be able to pass a fertility examination at the animal clinic of a major eastern veterinarian school. Collection of this horse's semen specimen did not put the purchasers' minds to rest, as the evaluation showed the horse to have marginal fertility.

It was suggested, however, that this marginal fertility might be attributed to the exogenous steroids prescribed for the stallion; it was said that this calmed the fears of the buyers and they didn't ask for their money back.

Secretariat was test bred to an Appaloosa mare, who promptly conceived and subsequently foaled a spotted stud colt. Of course, Secretariat went on to win the Triple Crown and to breed many of the best Thoroughbred mares in the world until his untimely death from founder (fever and dialation of blood vessels in the hoof).

But not all cases of low fertility have such a positive conclusion. Assault also won the Triple Crown but produced very few offspring due to poor fertility.

Another part of a fertility evaluation that is frequently overlooked is a measurement of testicular size, since evidence suggests a correlation between testicular size and sperm output. In a mating system where natural cover is the method of depositing the sperm cells in the mare's reproductive tract, minor variations in testis size and semen production do not seriously alter the outcome. Similarly, penis length and size of the glans penis would have little or no influence on pregnancy in the breeding system based on artificial insemination.

However, I believe that penis length is somewhat correlated to the fertility of a stallion under a hand-mating routine. A stallion with an abnormally small, short penis is unable to seat the glans penis against the cervix of the mare. It is this coupling of the concave glans penis with the mare's cervix that allows for the sperm to be deposited in the most advantageous location for the survival and transport to the site of fertilization in the upper oviduct. If this coupling does not occur, much of the semen is deposited in the vagina, where it can and does get expelled when the mare squats to urinate. Semen projected through the cervix into the uterus is not so discharged. These facts lend credence to an observation made by Burl Hill, an old stud manager from whom I absorbed a lot of knowledge. One afternoon Bob Gray, former author, editor, and publisher of Cordovan Publishing, came to interview this expert for an article in his *Horseman* magazine.

"What do you think is the single most important factor in determining a stallion's fertility?" Bob asked. "And please try and phrase your reply so that I can quote you."

Burl responded by placing his hands about two feet apart as though he were demonstrating the size of a fish in a big fish tale. "A big organ," he replied.

I've often wondered how Bob translated that response into the written word.

By now you may be wary of ever owning a stallion. Be assured that at least 90 percent of all mature males have acceptable fertility. The other 10 percent that are marginal or infertile are cause for alarm when large sums of money are spent in hopes of recovering the initial expense plus a profit by standing the stallion at stud. It is from this perspective that we must decide whether the expense incurred by performing any or all of these fertility tests is justified.

CHAPTER THIRTEEN

SEMEN OUTPUT

As director of research and stallion management for Horse Breeders Service, a California company that shipped stallion semen worldwide, my responsibilities included collecting semen from fifty stallions. One day I received a call to drive several hours away to collect an ejaculate from a large Belgian draft horse stallion. I was twenty-five years old and almost all of my experience had been in collecting light horse breeds such as Quarter Horses and Thoroughbreds. Being young and naive, I did not hesitate, as I thought that there would not be much difference in collecting semen from a slightly larger horse.

The ranch office was located in a eucalyptus grove surrounded by a fence whose rails resembled horizontal telephone poles. Initially, I remember thinking that the fencing design was a bit of overkill for restraining animals smaller than elephants.

The farm manager met me, and as he showed me around, I soon became intensely aware that all horses are not created equal. These horses were some of the largest in the world, standing twenty hands (six foot eight at the withers) and weighing close to two tons (that's four thousand pounds!).

Beads of cold sweat formed on my forehead as I thought about having to get underneath the belly of one of those frenzied Goliaths as he was balanced on his hind legs with a full erection. I was sure that the gentle giant would not be tremendously concerned about stomping on me. My anxiety was further heightened as the stallion manager related the story of how he had been showing this particular stallion at the Cow Palace in San Francisco, when the horse accidentally placed his hoof on top of his foot, breaking every bone.

At the collection area, a huge mare was tied to the fence (which now appeared to be appropriately built). I was assured that the jump mare was very gentle and would willingly cooperate for the procedure. We then parted company to prepare for the event.

As the last drop of hot water entered the artificial vagina, I heard someone shout, "Are you ready?" I glanced up to see a mountain of a horse coming around the corner of the barn with the stud manager attached to the lead shank. I realized that the diminutive fellow (about five foot four and 125 pounds) was not touching the ground with his feet. The stallion's head was raised to about ten feet in the air and the handler's hand was a couple of feet down the short lead shank. Stretched out with his arm fully extended over his head, the man was about seven feet from his fist to the tip of his toes. Add to this the two feet of shank between his hand and the stallion's head and the total comes to nine feet—one foot short of reaching the ground. This sight greatly concerned me since there was no way that the handler could assert any regulatory control over the progress of the mammoth stallion toward the hobbled mare.

In a panic, I screwed the plug into the AV and ran to the side of the mare just as the stallion reared to mount. As he thrust forward, I somehow managed to deviate his penis into the AV. But my problems were just beginning. This particular pile-driving two tons of flesh was a dancer! As he thrust to breed the mare, he would rapidly alternate his weight—first on one hind foot, then the other. While few stallions perform this breeding dance, it is an interesting behavior unless you happen to be standing at the rear of a mare trying to collect your partner. Then, each "cha cha cha" dance step brings four thousand pounds of weight to bear upon the ground and it becomes extremely important not to get out of sync with the one in the lead.

And, not to be overlooked is the fact that you are standing beside the hip of an equally huge mare who is not only bearing her weight but a proportion of the weight of the stallion upon her back. It is normal for a mare so disposed to balance this weight by spreading her legs apart in a breeding stance as the stallion mounts. So you must be sure that as the mare picks up her hind foot to make this wider stance, she does not put it down on your toe, as you are in the perfect position for such an occurrence.

Suddenly, the stallion went rigid and ceased stomping. He was ejaculating and I had survived! The trauma was over. At this moment of relief, I felt a warm sensation on my hip and leg. Peering to see what was causing this unusual feeling, my stomach sank when I realized that, in the frenzy of the event, I had forgotten to put the collection bottle on the AV. Semen was pouring into my lab coat pocket—straining through the white cotton before flowing down my leg.

Not wanting to have to spend the whole day for naught, we waited ten minutes and repeated the process. This time the stallion was much more orderly and I, of course, remembered to attach the collection container. The Belgian stallion obliged by producing enough semen to fill the 250 cubic centimeter baby bottle twice—a fact that amazed me and piqued my curiosity about the differences that might have existed between the ejaculate in my pocket and the one in the graduated cylinder.

Since then there has been much discussion and some research on the effects of "doubling" a stallion. Much of this information seems to be contradictory, but the confusion appears to be centered around a clear definition of doubling, particularly a standardization of time intervals between ejaculations. For example, in one case study I collected a stallion twice a day for fourteen days; once in the morning and again eight hours later. Using standard laboratory evaluations for semen fertility, there appeared to be no differences between the two ejaculates of the same day. The indication was that doubling a stallion for a limited period of time had no effect on semen volume or quality. About the same time, a study was released from Colorado State University stating that B.W. Pickett and associates had completed research on stallion doubling that indicated a reduced volume and concentration in the second ejaculate. A close reading of the research project showed that the second ejaculate was collected ten minutes after the first. So instead of the two projects indicating contradictory data, it should be noted that the time between doublings seems to be critical in maintaining the quality of the second ejaculate.

These results were further explained by some enlightening work done on the prediction of daily sperm output in stallions by Dr. James Dinger at the University of Connecticut. He used ten thoroughbred stallions to determine the average daily sperm production (DSP) and the effect multiple collections had upon fertility. Spermatozoa are produced by the stallion's testes on a continual basis; these sperm cells are primarily stored in the tail of the epididymis awaiting ejaculation and are referred to as the extra-gonadal reserves. A stallion that is bred daily will exhaust his extra-gonadal reserves in five or six days; stallions bred two or three times a day will exhaust their reserves in two to three days. Once these reserves are used up, a stallion can only ejaculate as many sperm cells as he produces between collections or breedings.

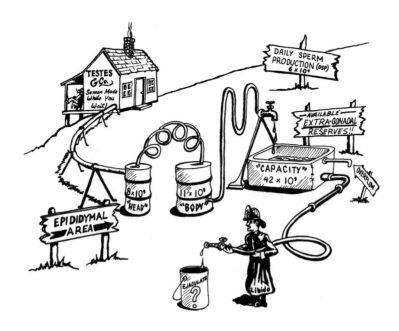

Semen production. Drawing by Steve Eaves. Concept by Jim Dinger.

The mean daily semen output of these mature stallions was 6.6×10^9 (that's 6.6 billion sperm cells). Since one billion cells is the accepted standard for a satisfactory breeding, these stallions theoretically produced more than enough spermatozoa to hand-breed three mares per day as long as the covers were spaced far enough apart.

The influence of this time interval can be seen in the following chart, which shows the relationship between time between ejaculations and the amount of the ejaculate in stallions with exhausted extra-gonadal reserves.

Time Between Ejaculations (in hours)	Number of Sperm (in billions)
24	6.6
12	3.3
6	2.2
4	1.4
2	0.6
1	0.3

This information becomes extremely useful for stud farms dealing with horses whose breed associations do not allow artificial insemination. When trying to maximize the number of mares bred during the breeding season, the number of live covers a day per stallion becomes a critical economic factor. When sixty mares are booked to a horse (live cover), the peak of the breeding season can become a nightmare for a stallion who can't handle being doubled or tripled. It is up to management to make this process as easy on the stallion as possible and, at the same time, give every mare the best possible chance of conceiving. As research has shown, spacing out the covers over the course of the day usually does not result in a significant reduction in semen quality.

The biggest problem in doubling is sex drive. Some stallions who are being doubled or tripled will gradually seem to lose interest and refuse to cover a mare. The sooner that the people in the breeding shed recognize a potential problem and try to adjust the horse's breeding schedule to his individual libido the better. Scheduling a stallion's breeding book is a very important job done best by someone who understands not only the physiology of the stallion's reproductive organs but also the psychology of that particular horse's sexual behavior.

When sex drive falls below what is expected due to either age or overuse, productivity techniques are sought that will enhance sexual activity. For many years, testosterone has been injected to increase sex drive. Unfortunately, the results have been unpredictable and often disappointing. Recent attempts to correct this varied response to the injection have included the use of an indwelling cannula—on the order of an IV hookup through which a patient can receive various fluids. In this case, the fluid contains a gonadotropic-releasing hormone (GnRH), which is released at physiological levels in a continuous fashion in hopes of mimicking the body's natural production level.

In the few cases of diminished libido in stallions that I have had to deal with, I have used 10,000 IU of chorionic gonadotropin to get a more natural physiological dosage of testosterone from the poorly performing male. Chorionic gonadotropin is high in ICSH, which is the hormone that stimulates the testes to produce testosterone. By going back one step and injecting ICSH, the response of increased sex drive seems to be stronger in some cases than that produced by using testosterone.

Of course, no injection was necessary for that two-ton Belgian dancer who moved just as feverishly during the second mount ten minutes after the first. But that was just one day!

RULES OF THUMB

There is a large variation in the breeding performance potential of stallions, especially between the ages of five and twenty. Care should also be taken not to overbreed the young horse as it may cause a permanent decrease in sperm production.

Age of Stallion	Number of Mares to Be Bred Per Year	Frequency of Service
2	10–15	2–3/week
3	15–30	1/day
4	24–40	2/day
5–19	40–60	2/day
20 and older	20–40	2/day

PART III

CARE AND
MAINTENANCE OF
THE STALLION

CHAPTER FOURTEEN

STALLION FACILITIES

Once upon a time a student of mine told me he knew where there was a stallion for sale that he was sure I would like and that was reasonably priced. This sounded like a fairy tale, so it took some coaxing before I agreed to drive some sixty miles to look at this horse. Upon arriving at a little farm somewhere in the Maryland countryside, I was ushered behind the house toward an outbuilding that looked like an old-fashioned chicken house with a net wire chicken yard—all of which was in disrepair.

I knew it, I thought to myself, I will have wasted nearly half a day driving to and from this place to look at a stallion prospect living in a chicken coop. At any moment, I expected a small, half-starved horse with crooked legs to emerge. Instead, as we approached, a beautiful 16.1-hand chestnut stallion with a blaze face and stocking legs began to emerge from the dilapidated building. So striking was this horse that his appearance took on a mythological quality, as if the chicken house were giving birth to one of the most appealing animals God ever created. Once he was out of the coop, it seemed impossible to imagine fitting such a magnificent animal back into those tight quarters. The student had been correct. I was impressed with the animal and before I went to bed that night, I owned that stallion.

The conclusion to this story, however, is the exception rather than the rule. I have looked at hundreds of stallions in similar circumstances and rarely does one find a diamond in a muck heap. The horse industry exists primarily because of the beauty we associate with horses. To paraphrase Winston Churchill, there is something about the outside of a horse that is good for the inside of a human. This is especially true when that horse is well turned out—fit and groomed. Add to this picture well-kept and attractive surroundings and the economic value of the horse begins to come to mind: The key image here is money.

This horse booked five mares standing in a chicken yard and seventy mares after being moved to a more traditional setting.

This concept certainly worked for the chicken coop stallion. Housed in a chicken yard, this extremely handsome horse bred five mares. After a mildly vigorous advertising campaign and relocation to an adequate horse facility, the same horse bred seventy mares over the next two years. This points out that no matter how good the product, packaging and marketing play a key role in consumer acceptance: especially in an industry whose foundation is almost entirely based on our romance with horses.

As horsemen we all have our own goals involving horses. The images of our success are reflected in our mind's eye: winning a major event, breeding a superior horse, riding into the sunset with the partner of our dreams. It is this imagery that sells horses and horse products—as well as laundry detergent, perfume, coffee, and a host of other products that Madison Avenue can plug into this dream. All of this should be kept in the forefront of your mind during the designing of facilities for your stallion or stallions.

BARNS

Barn building materials generally reflect locally available resources and conform to traditions proven to create "the right look." Facilities can be personalized to include advertising signs, accesses enhanced by landscaping, and unique architectural design of buildings. While these elements may cost a great deal of money, paramount in all of this is the care taken to maintain the facility in a neat, presentable, and functional fashion.

There are a few basic concepts included in most barn designs. It is generally accepted that the minimum size for a

One of the four-stall stallion barns at Gainesway Farm, Lexington, Kentucky.

light horse stallion's stall is twelve by twelve feet, or 144 square feet. Many barns enlarge this stall size because of the time spent in it by the occupant. I am not sure this is an advantage, because oftentimes a larger stall promotes more physical activity than is safe in a confined area. Again, the individuality of the horse living in the larger quarters plays the major role in determining the success of this design.

Another factor to consider in providing a healthy environment for the housed stallion is ventilation. Stalls and barns should be free of drafts. Wind currents produced by air sucked through a narrow opening act like a tiny jet stream. This kind of air movement gives rise to colds and flu. On the other hand, the exchange of large volumes of air that generally replaces damp, stagnant, or dusty air with clean, fresh air proves to be a health-giving situation. Tobacco barns, designed for drying the leaf through the natural movement of air currents, have been said to convert into excellent horse facilities.

Back door ventilation.

STALLION PADDOCKS

Paddocks also change in appearance and construction according to geographic location. In the Southwest, the fencing may be pipe and steel rods. In Kentucky, the materials are likely to

New tobacco barn construction designed for maximum ventilation at Margaux Stud, Lexington, Kentucky.

be four-board wood fences; in Florida, mesh or woven wire seems to be the preference. All these fencing materials have their advantages and disadvantages, but the key to paddock design is that the fence be tall enough, strong enough, and safe enough for your purposes. As the old-timer's saying goes, "Horse high, bull stout, and hog tight." More specifically, horse high generally means four feet or better; bull stout measures the strength of the building materials; and hog tight means placing a post every eight feet with little opportunity or space for a horse or a hog to reach through the fence.

Personal preference plays an important role in determining the size of the paddock. It is my opinion that a stallion paddock should be large enough to allow the horse free movement up to and including performing a strong gallop. The minimum size, therefore, should be a two-acre paddock with rounded corners. An area of this size also reduces the stress placed on the turf by the horse's excrement and heavy trafficking, allowing grass to prosper and fecal matter to degrade.

Traditional four-board fencing surrounding stallion paddock.

BREEDING FACILITIES

It is certainly nice for employees and horses to have a breeding shed that is at least partially protected from the elements. The minimum size should be thirty by thirty feet, with twelve-foot-high ceilings. Actually, in this case, bigger is better. In California, I visited several farms with sixty-foot

Breeding shed, Spendthrift Farm, Lexington, Kentucky.

circular breeding rooms that I thought were very functional; without corners, you are unlikely to get pinned by an unruly animal.

A laboratory attached to the breeding facilities is a must for any well-equipped breeding farm. While it doesn't need to be fancy, hot and cold running water, sinks, countertops, and electrical outlets go a long way toward supporting the refrigeration and incubation equipment that may be necessary for the proper management of semen. All of this can be contained in a room about the size of an average bathroom; that is, unless the management requires research work or semen is to be handled by more than one or two people.

Looking at all the wonderful equipment and facilities that some farms possess lends a certain credibility to the quality of care and expertise that is to be expected. But let us not forget that horses managed to propagate quite successfully out on the open range with very little intervention by man. When designing a stallion operation, it is a good idea to keep in perspective the goals and realistic expectations attainable for your personal operation and battery of stallions. If economics are a consideration, remember the value of the stallions and their prospective income will play a major role in determining what type of facility is likely to be supported by their efforts.

Interior of padded breeding room.

Padded hot water heater in breeding room.

Before breeding, the stallion gets cleaned in a special area with a drainage system.

NUTRITION

Behaviorists often say that the desire for food is stronger than the desire for sex. This is a half-truth; the desire for food is only stronger than the desire for sex when you are starving—at least it's that way for stallions. And it is our job as managers of breeding stallions to be sure that our horses are not forced into having to make a choice. We want to continuously provide all the necessary nutrients so that the stallion is inclined to choose sex every time it is offered and his fertility is not hampered by any lack of nutrition. A proper balance of nutrients is the basis of the true aphrodisiac.

Recognizing the importance of nutrition to reproductive productivity invariably brings up the following question: How do the nutritional requirements change for the sexually active male? Unfortunately, there is no concise answer to this question because every stallion is an individual and variation among horses is great. Still, finding the correct answer for your stallions is paramount to maintaining sound management practices and sometimes the conclusions may even surprise you.

I began to appreciate the uniqueness of individualized nutritional needs when my continued exposure to higher education conditioned an attitude toward testing and measuring. One breeding season my research animal became a mature Quarter Horse stallion with a book of thirty mares that had to be bred in ninety days. After calculating a well-balanced diet that was to be fed at a constant amount over the course of the breeding season, I made plans to record his weight every seven days.

The first week of breeding season went exceptionally well; the horse covered a mare every single day. At the end of the week, I weighed the horse and was very surprised to learn that the stallion had gained weight. To my astonishment, this trend continued, and by the end of the ninety days the horse had gained nearly one hundred pounds. Looking back at the behavior of this stallion during breeding season, it was easy to see that he became more sedentary as he settled into the routine of the breeding season. He was caught and bred at the same time each day and I usually found him in his stall, quietly waiting. His behavior was very different during the off-season. The stallion spent most of the time out in his paddock running and cavorting around—obviously burning more ATPs (energy) and needing more feed to maintain his weight.

In another instance, my research animal was an older Thoroughbred stallion on whom breeding seemed to have the opposite effect. This horse was fairly quiet and manageable during the nonbreeding season, but once mares started to come in to be serviced he became increasing rowdy and nervous, constantly pacing his stall or running the fences of his paddock.

During the off-season, this stallion maintained a relatively constant weight on ten pounds of grain and fifteen pounds of hay. During the active breeding season, doubling the daily grain ration to twenty pounds still did not curtail the weight loss. Only the end of the breeding activity fixed that.

Even though these two horses represent extremes, they illustrate that the nutrition necessary for breeding and sperm production can be secondary to the nutritional needs determined by the way the stallion adjusts his lifestyle to the breeding season—the physical and mental stress that he puts himself through. This fact holds true even in a more natural environment. Some stallions turned out with a band of mares present a very tranquil scene. As long as no other male tries to interfere with their herd, stallions such as these are content to graze along, breeding the mares as they become receptive and letting the boss mare handle herd politics. On the other hand, another stallion may be constantly searching for trouble, disciplining mares and worrying about intruders. It is obvious that the nutritional needs for these horses with such diverse dispositions would be different. With these diversities in mind, let's discuss the basic nutritional components that might be manipulated in order to create diets to fit the individualized needs of various types of stallions.

ENERGY

The main consideration in maintaining homeostasis for a hyperactive breeding machine is energy. There are two ways to provide the overly active stallion with more energy so that he won't loose too much weight during the breeding season.

The first choice is usually to increase the amount of grain that is fed. Since the grain portion of the diet has more energy per pound than hay, this increase will result in higher energy totals for the day.

The limiting factor with this method is the horse, himself. The average horse will eat somewhere between twenty-five to thirty pounds a day; for safe digestion, at least ten pounds of this total feed consumption should be roughage (hay). This sets the maximum grain ration at twenty pounds a day.

For horses that still loose weight and can't be safely pushed to eat any more, a second option is available: *Increase the amount of energy per pound of grain.* There are two common ways to achieve this goal:

1. Add more high-energy grains like corn and barley to the grain mixture, or
2. Supplement with fats or oils up to 10 percent of the grain ration weight.

Historically, flaxseed or linseed meal were the supplements of choice to provide glossy hair coats and better condition. While both of these additives do have a fairly high energy content, they are relatively expensive. Today there are cheaper, if not better, sources of energy available, such as corn oil and fats, which have been treated so that they are not as likely to become rancid and therefore are more palatable.

VITAMINS

Wheat germ oil has long been a favored supplement for breeding stallions since it supplies both additional energy and contains a large amount of vitamin E—the vitamin that has received a lot of press because of its association with improved sexual performance. Actually, in horses not suffering from a deficiency of vitamin A or vitamin E, there does not appear to be any increase in reproductive performance from the overfeeding of vitamin E, either from wheat germ oil or from other sources (megavitamin therapy).

Adequate dietary amounts of vitamin A and vitamin E are necessary for the maintenance of the tissues in the reproductive system. These requirements are supplied in most standard horse rations and usually do not require supplementation. This is also true of other vitamins such as the B complex water-soluble vitamins associated with protein synthesis and blood building.

PROTEIN

Protein is an important part of any ration. Having an adequate supply of the essential amino acids is as important for the breeding stallion as it is for the competitive athlete. While linseed meal and flaxseed are also recognized as concentrated sources of protein, they do not provide *high-quality* proteins—that is, they are not good sources of the type of proteins normally lacking in the diet of the horse. Today, soybean oil is acknowledged as the protein supplement of choice. It is an inexpensive and high-quality source of concentrated protein.

Determining an adequate level of protein for the breeding stallion is related to his age and activity. The younger, more active stallion needs a higher level of protein than the more mature, less active one. For stallions eating ten pounds of hay (containing 10 percent protein), it is reasonable to expect that their total protein requirement would be met by feeding a good 12 to 14 percent protein grain ration.

Feeding the stallion during breeding season boils down to the same good nutritional management considerations that apply to the feeding of horses year-round: a proper level of energy, an adequate amount of quality protein, mineral, and vitamin concentrations that meet the requirements of the species, and plenty of fresh water. Since animals do vary, the requirements for these nutrients also change, and it takes the eyes of a good horseman to adjust the ingredients to satisfy varying needs.

And, yes, good horsemen see things differently. Some stud managers feel that a stallion should go into breeding season slightly overweight to compensate for an expected weight loss due to the increased activity. This is just another way of supplying energy to the active stallion.

Other horsemen feel that a stallion should be maintained at a slightly overweight condition before, during, and after breeding season because they are selling the services of their stallion that is based, in part, upon his appearance. To most livestock people, a little fat is beautiful and this image makes the horse more marketable.

Then, too, there are those horse persons who believe that the trim, fit animal is healthier and the health of the horse is the most important consideration. These caretakers maintain that the diet should only replace the energy being used and should not cause fat to be laid down.

The fact is, all these opinions can be correct. It is simply a matter of which management system is in operation and what the farm's priorities are. In actuality, under the care of conscientious horsemen who supply a balanced diet to their charges, a stallion's reproductive productivity is not likely to be hindered or enhanced by nutrition except, perhaps, in the case of the older male.

Maintaining stallions that are beyond middle age oftentimes requires adjustments to prolong longevity and fertility. Horses, like people, show a great deal of individual variation in the aging process—the mechanisms controlled in large part by genetics and influenced by environment. Even though we cannot control the genes, we can sway them somewhat through manipulation of the environment. Stress reduction can be accomplished through identification of factors that seem to cause discomfort to specific stallions. Heat, cold, too much rain, or too little rain are simple stressful situations that can be modified through the use of fans, heaters, blankets, and stalls.

Advancing age also affects nutritional factors in a variety of ways that contribute to common geriatric conditions. For example, constipation is a familiar problem in human aging, but in the horse, it can lead to major colic problems that may result in surgery or even death. As with humans, fiber in the diet is one of the more common preventative measures. A 20 percent bran ration or feeding hot bran mashes helps to improve movement in the gastrointestinal tract.

Coupled with constipation is the inability to chew coarse feeds. A stallion in his declining years (somewhere between eighteen and twenty-eight) will often develop gaps between his molars. Add to this the worn-down surfaces of very short teeth and the older horse has trouble chewing and swallowing those feeds usually given him. These problems can be overcome by either feeding a pelleted ration or soaking the grain ration in water. Pelleted feeds require very little chewing and soaked whole grains are also easier to swallow—both factors that will ultimately increase the available nutrients for the older horse.

Although little is understood about the maintenance of epithelial tissue, it is another area of geriatric concern. Skin and the reproductive tract are both made of epithelial tissue. The same minerals, vitamins, and proteins that make vivacious, healthy-looking skin contribute to maintaining a productive reproductive tissue. Both vitamins A and E have long been known to be important biological catalysts in this process. Normally, the horse's body makes vitamin A from a closely related precursor, carotene. However, in some older animals, the ability to either make or use this vitamin becomes somewhat impaired. Vitamin injections or oral supplementation (e.g., cod liver oil and wheat germ oil) may offset this problem.

CHAPTER SIXTEEN

OFF-SEASON MANAGEMENT

After midsummer, the stallion management team as well as the rest of the breeding shed personnel are ready to give it a rest. Ideally, this break will be well deserved due to the hectic pace of a successful breeding season. When the season began in late winter, peaked in the spring, and tapered off in the early summer, those involved were eager to put last year's breeding season behind them. But rest is not in the cards for the weary stallion manager, who must now begin to make plans for next year's breeding season. The successful management of the stallion during the off-season is essential to a successful stud career from both a business and a management point of view.

The culmination of the mare service part of the season is an ideal time to begin off-season management by evaluating the stallion's fertility. With every horse having individual variation in semen production and sex drive, this is an appropriate time to measure the effect that the previous season's breeding schedule had on sperm output so that assessments can be made about the capabilities of the horse for next year. Cultures done on semen and the urethra will also allow ample time for treatment should the horse have picked up a pathogen during the recent sexual activity.

If all is well after a comprehensive stallion reproductive examination and it appears that the stallion is not suffering from any breeding fatigue, one might consider the possibility of moving the horse to the opposite hemisphere for a second breeding season in the same year. On the other hand, evaluations that show reproductive activity has depleted semen reserves or that libido is waning are symptomatic of a horse that needs some rest and recuperation. Whether this double hemisphere duty is practical or not, the summer is the opportune time to prepare the promotion campaign for next winter and spring. Advertising deadlines are usually several months in advance of publication dates, so summer is the time to make decisions about how much to spend and where to spend it.

It is also a good time to hire a photographer and set up a date for photo sessions when the stallions are slick and shiny due to the warm weather. Also, higher quality photos are more likely because of the light, and trees and flowers will help produce more aesthetically appealing pictures.

Other summertime activities should include assessing the success of the previous season and what adjustments might enhance the upcoming activities.

SOME TOPICS TO CONSIDER

Should the Stallion Remain at the Same Location?

Stud horses tend to be like famous people. When a person is far enough away from home and has a degree of fame, the public wants to meet him. In his own backyard, where everyone has become familiar with his presence, there is no urgency to make his acquaintance. This phenomenon seriously affects stallions.

A stud that has been in one location for several years tends to lose some of his appeal unless he is one of those rare individuals capable of producing babies that are ten times better than their mamas. Amazing enough, even if he is, too often the fickle nature of mare owners dictates that they are always off looking for the "new" superstud that can make a silk purse from a sow's ear. Very often, interest in a stallion drawing a dwindling number of mares may be rekindled simply by moving him into a new area with different breeders.

Should the Stud Fee Be Raised or Lowered?

Determining the stud fee is a very difficult marketing problem. It's an issue that requires a great deal of knowledge about the market and the breeders for which the stallion is competing. Generally speaking, pricing the stud fee of a stallion below his market value will increase the book of mares coming to his court. However, consideration must be given to devaluing his offspring already on the ground as well as the stallion's image or profile.

Should the Amount Spent on Advertising Be Raised or Lowered?

Deciding how much to spend on advertising requires a thorough analysis of the cost effectiveness of the previous dollars spent in each advertising market. The rule of thumb is to spend approximately the equivalent of two stud fees. As with all rules of thumb, variations can be considerable. The amount spent will depend on the methods used and the size of the target market. Some of the methods are direct mail, trade or breed journals, hand bills and posters, etc.

Should Breeding Incentives Be Offered to Mare Owners?

There are also many ways to offer incentives to make breeding to a particular stallion more attractive. These enticements will work to increase the number of mares bred. Examples of inducements include giving a reduced stud fee to an owner for booking a number of mares; reducing the stud fee to owners of mares that are nationally recognized performers or that have produced outstanding offspring; and offering a cash reward to the first foal by a stallion to win a championship. Creativity in marketing will make a difference—a difference limited only by your own imagination.

Should the Stallion Be Advertised During the Off-Season?

Questions need to be addressed to determine whether it is important to keep the stallion's name and reputation on the consumer's mind during the summer and fall when many of his offspring are going to be marketed. Even though not directly related to stallion service promotion, this image making can have a substantial effect in supporting a horse over time.

All these questions concerning the profitability of a stallion's breeding life should be answered early enough, before next year's contract, to make a difference. I have found the best time to do this is when you are counting up the costs and credits from the previous year's business. The stallion owner is obliged to face the realities of profit and loss and recognize that hard business decisions must be made if profit is to be an important consideration.

Along with all the CEO decision making about financial matters such as image making, mare portfolio, and star-quality longevity, more mundane matters must also be addressed—things such as feed, exercise, and herd health. Exercise is easy to neglect when a horse is not going to the breeding shed every day, yet we know that exercise is necessary for longevity and good health. Feed is always important to health and is an influential part of rest and recuperation programs. If weight loss has been the result of breeding activity, energy in the diet should gain back those lost pounds. If weight was maintained during the breeding activity, you may have to cut back on calories until a new homeostasis is established.

Periodic observation and cleaning may be necessary during the off-season.

Herd health during the off-season is more often neglected than feed adjustments, but it too is necessary. When sexual activity subsides, debris such as sheath smegma rapidly builds up in and around the foreskin, or prepuce. This buildup is natural in a seasonal breeder like the horse, but it can cause problems if not dealt with. Formation of beans will cause discomfort and, in some cases, lesions of the penis. All this can be controlled by periodic observation and, when necessary, cleansing of the genitalia. Other herd health measures should, of course, include parasite control, vaccinations, and dental care during these nonbreeding months.

Clearly, the off-season in the stallion game is far from inactive. As in any operation, much planning and work must go on behind the scenes to have a successful opening and profitable seasonal run.

PART IV

THE MARKETPLACE

THE THOROUGHBRED OR RACING-TYPE STALLION

Even though this chapter and the one following contain similar information, the emphasis and focus are different, reflecting the fundamental differences in the ways the horses will be put to use. It is hoped that this treatment will serve as a guide to horsemen who approach the horse business from a variety of perspectives.

SELECTING A STALLION PROSPECT

There are many reasons why people decide to add a stallion to their farm portfolio. Perhaps they own a horse farm with the facilities and labor force adequate to stand a stallion. Or perhaps they own one or more mares and wish to own or share in the ownership of a breeding stallion. Or perhaps they temporarily have lost their sanity.

Fortunately, the marketplace usually has an abundance of stallion prospects available at a wide range of prices. Over the past thirty years, I have stood more than three hundred stallions and, unfortunately, regardless of their initial price tags, most of them didn't make money. Since the IRS assumes that we are all in the horse business to make money, let's examine the advice given to everyone looking to purchase a stallion prospect.

A male horse worthy of a harem of mares needs to be superlative in pedigree, conformation, performance, and maybe disposition. This is a general statement and therefore subject to a variety of interpretations. It is my purpose to define more specifically what this statement means in terms of standing a stallion to the public's mares in order to make a profit.

Let's start with the word *superlative*. We learned to use the adjectives *good, better,* and *best* to show degree of comparison. *Best* is the superlative, of course. But for purposes of our discussion, what exactly are we comparing? Are we thinking in terms of the best of all the other horses in the world or the best of all the other horses in the same barn?

The answer to this question is determined by the population of mares that the stallion intends to draw upon. If there are a thousand mares in a geographical area that would be worthy of going to a stud within a given stud fee range, and your stallion is the best in that area in that stud fee range, you could reasonably expect to get a significant proportion of those mares.

Knowing the competition for those mares provides the comparison for who is best, but knowing the population numbers and quality of mares you are competing for helps to determine that range of stud fee.

I believe that a stallion's stud fee should not be more than

50 percent of the value of the mare you are trying to attract. For example, if those thousand mares have an average value of $10,000, the horse you are offering should not have a service fee over $5,000.

This rule of thumb is also relevant in determining exactly what group of stallions a horse is competing against. In this case, any stallion with a stud fee over $10,000 should not be competition for this particular population of mares (mares worth $10,000).

With his competition pinpointed, does the young male horse entering stud have to be the best in all of the areas listed: pedigree, conformation, performance, and maybe disposition?

It would be great if he were, and it would be very easy to fill his book, *but* unfortunately, only a few horses can attain this unique position even within a specific economic stratification of mares. On the other hand, the stallions that are actually blessed with these superlative qualifications leave room for other horses who are less than "the best" in all areas. This leads us to the task of determining how to rank a horse in relation to the big four.

To state the obvious, a horse superlative in three categories is better than a stallion highly qualified in two. A horse that excels in one category might still have a reasonable chance of being stallion material if he is promoted to the right group of mares and is in competition with the right group of stallions—as long as his glowing best quality is not disposition.

Disposition is not an easy trait to market as an outstanding reason to breed to a particular horse. Undeniably, this trait does have economic importance for those handling the stallion on a daily basis, but most mare owners feel that the mare contributes more to the disposition of the offspring than the stallion.

Also, marketing disposition is difficult because of the wide range of attitudes held by mare owners. The desirable temperament for one horseman may be quite different from that of another. A docile horse may be a joy to work around, but a more spirited, aggressive animal may sire better performers. Horses like Native Dancer and Nasrullah were definitely not pliable individuals willing to perform whatever task man might wish, yet they were outstanding as sires. Therefore, excellence of disposition, like beauty, tends to be in the eye of the beholder. There does not seem to be a consensus of opinion that describes the truly superlative disposition for a racehorse.

Nasrullah at Claiborne Farm. Courtesy of Keeneland Library, Lexington, Kentucky.

Now three areas remain for a stallion prospect to reach the superlative: pedigree, performance, and conformation. It is always possible to breed mares to a beautifully conformed horse. This does not mean that differences of opinion do not exist on ideal type, but almost all horsemen can agree on

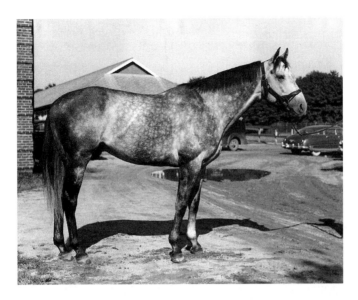

Native Dancer at Belmont Park. Courtesy of Keeneland-Morgan, Keeneland Library, Lexington, Kentucky.

those horses that have attained the pinnacle of conformational excellence. If this is the only superlative of the stallion prospect, however, he is likely to be doomed to a low service fee.

In the early 1970s, I stood such a horse. Everyone thought him a handsome specimen, even though he had chipped both knees as a two-year-old, a fact that contributed to his poor performance record. His pedigree, while not unknown, also lacked luster. The only superlative this stallion possessed was his good looks and his rock-bottom purchase price.

An analysis of mares that might be bred to this horse showed a large population of hunter and jumper mares. The horse was introduced at a $400 stud fee and immediately booked fifty mares. After his foals were born and went into in-hand competition as yearlings, they won more than their share of trophies and ribbons (a poor substitute for purse money). This led the owners to believe that the stud fee should be raised a few hundred dollars—a move that reduced the horse's book to fewer than twenty mares. Returning the stud fee to its marketable price, the stallion's book went back up. The horse finished out his stud career standing as a conformation sire for the same low fee.

A superlative in performance is another story. A horse that has shown brilliance in either speed or class magnetizes mare owners. Having large dollar figures and capital letters, especially G's (as in G1, G2, or G3) under the performance record, is impressive to breeders, but it is not totally necessary. Raise A Native was a champion two-year-old colt with impressive victories and only $46,000 in purse money. On the other hand, champion Carry Back, racing's first millionaire, needed an outstanding list of racing statistics to attract high-quality mares.

The limits to the benefits of a superlative in performance was illustrated to me several years ago when I leased a horse that had won almost a quarter of a million dollars, had placed in the classics, and from his first small foal crop, had produced a superior runner. Surely, I thought, this horse had enough performance to stand at stud in the South. Yet, at a reasonable stud fee ($1,500) for the times (1982) and in an area with a great deal of breeding activity in this low-price range (Louisiana), this horse never serviced more than fifteen mares a year. The problem was pedigree.

An old saying among stallion owners puts it bluntly, "When talking to the general horse public, you say who your stallion is by, and if you have to say any more to identify that sire, you don't have a stallion." To carry this one step further, in order to excel in pedigree for marketing shares or stud fees, a horse must be by one of the fashionable sire lines of the period. Historically, this will limit the top half of the pedigree to very few individuals, while the bottom half is open to a wider range of individual names (although the number of females capable of producing fashionably bred sons is also a rarefied group).

When considering superlatives, it is necessary to rank the evaluation categories in order of priority.

1. Pedigree
2. Performance
3. Conformation
4. Disposition

And this leads to a restatement of my original advice. In order to have a reasonable chance of being a commercial success, a stallion must be superlative in at least one of four areas (pedigree, performance, conformation, and disposition) for the population of mares he is likely to attract. If he is excellent in only one of the first three areas, he certainly needs to be at least adequate in the others.

APPRAISING THE VALUE
OF AN UNPROVEN STALLION

It used to be that when a colt became a major money earner, he was syndicated for the amount roughly equal to his earnings. Things changed when leading money earner Nashua was syndicated for stud duty for much more than earnings. Then, Secretariat was syndicated while still racing—for more money than he could win if he raced several lifetimes. Seattle Slew did finish an outstanding racing career before he was syndicated for millions more than his racing income. He became a profitable investment when his services went for more than half a million per mare.

Lowered yearling prices are reducing stallion fees and stallion values. We are coming full circle as the dollar value of the Thoroughbred stallion again becomes related to his race earnings or the race earnings of his progeny. Even though not all the horses entering stud duty are leading money winners, their winnings are still a major part of the formula that determines breeding value.

The extent to which the amount won influences stud value depends on your point of view. Estimating the value of a stallion prospect also depends to a great extent on whether you are a buyer or a seller. Let's try and minimize these differences in favor of realism.

All Race Earnings Are Not Equal

Not every stallion who wins $100,000 is worth $100,000. A close examination of a stallion's racing record is necessary to determine his racing class, speed, distance, and precociousness—all traits that will affect his selling price. Weighing these factors will ultimately influence the appraised value.

In 1984 I was asked to appraise a horse whose particulars dramatically illustrate this point. This horse had earned more than $450,000 during a long and mildly illustrious racing career. Although his performance was impressive, he did not capture any graded stakes, and his pedigree could have been described as less than fashionable. His appraised value was $75,000.

Share-Value Appraisal

Another method used to determine a stallion's worth is to appraise his share value. This concept is best explained by an example: Let's assume that a limited partnership wants to purchase a stallion and divide him into forty shares. The big question for this group of investors now becomes: "What price will these shares bring in the marketplace?" Assuming that the shares will be purchased by mare owners who wish to breed to this particular stallion, it is reasonable to expect that a mare owner should be willing to pay two to five times the stud fee for a lifetime breeding to the horse.

Choosing a horse with the potential to book forty to fifty mares at a $5,000 stud fee should generate a share value of $15,000 and a purchase price of $600,000 (forty shares times $15,000).

Expenses-Income Method

We know that many of the horses bought and sold as stallion prospects cannot be categorized as either leading money earners or worth more than a half a million dollars in syndication. Evaluating the economics of such stallions takes on a more simplistic estimate of worth:

$$\text{income} - \text{expenses} \times \text{number of years to payout}$$

With an increasing number of stallion hopefuls in the United States standing for a $1,000 stud fee, let's develop a scenario to calculate their price tags. The first step is to examine the expenses.

If the cost of boarding the stallion is $600 a month, yearly board would equal $7,200. Add to this herd health and farrier fees—around another $800 a year—and the bare necessities total close to $8,000.

To this expense we must add advertising and management. For horses standing for under $5,000, two breeding fees would be a good estimate of promotional expenses. Similarly, two stud services should be allotted to a management personnel fee and one fee for insurance. This brings the total expenses to about $13,000 a year, or a little over $1,000 a month.

To begin to estimate the income of this or any stallion, an assumption has to be made as to the number of mares that will be bred. Depending on the position of the industry cycle and the geographical location, the statistical average number of mares per stallion is about ten.

This is an impractical estimate for a profit-making venture, so let's assume that the horse is going to be priced correctly and that he will breed forty mares, making the hypothetical income for our $1,000-per-service horse $40,000 a year.

Although this income seems possible, it isn't really realistic. Unfortunately, not all of these hypothetical forty mares will conceive and produce a live foal. While management expertise will play a major role in beating the national average, it would be wise to estimate that 50 percent of the mares will not produce live foals. According to *Thoroughbred Times* (September 13, 1991), the median percentage rates of live foals produced from 1989 breedings tied at 51 percent, 52 percent, and 53 percent. On the basis of these rates, the projected income will drop to $20,000.

A decision should also be made as to whether the stallion manager can offer breeding incentives to attract owners of quality mares in order to obtain a bigger and better book. It would be reasonable to expect that this management decision would reduce income by another 10 percent or, in this case, $2,000, causing the resulting income to drop to $18,000.

Also, a value should be factored into the formula to account for bad debts that are prevalent in the breeding business. Personal experience estimates that a 10 percent loss of income is reasonable to expect.

Let's look again at the calculation for income.

{(number of mares bred − number of barren mares) × stud fee} − 10% for bad debts and 10% for special deals

This brings the income for our example to $16,000. Add these figures to the original formula and we get the following information:

$$\$16,000 - \$13,000 = \$3,000 \times \text{number of years to payout}$$

If we assume that the net profit of the horse (in this case $3,000 a year) will be applied to the purchase price, how many years can you afford to pay for the horse? While the IRS allows a young breeding animal to be depreciated over seven years, most breeders would find it a financial disaster to stand a horse for seven years without making a profit.

In my experience, a reasonable amount of time for a horse to pay himself out is two and a half years. This number is based upon the industry practice of breeding to a young horse for three years (his honeymoon at stud). At the end of this period, a stallion must have produced a meritorious runner or breeders will move on to the next round of young stallions entering stud. Therefore, a young stallion in my barn must be able to pay himself off in two and a half years in order to be worth the financial risk of standing him at stud.

Using this two-and-a-half-year payoff time schedule, we now have sufficient information to complete the computation and estimate the value of this stallion based upon the original hypothesis. This example points out the validity of the expression, "It's hard to make money with cheap horses." Booking forty mares to a $1,000 stallion is a difficult prospect and the quality of the management will make the difference. This can easily be seen by changing a few values.

Suppose this horse produced 70 percent live foals instead of 50 percent. His income would become $22,400 and his estimated value would increase to $23,500.

Or suppose that he booked sixty mares with a 70 percent production rate and produced a $33,600 income, resulting in a changed profitable purchase price of $51,500.

Would you now be willing to spend $23,500 or $51,500 for the horse we initially appraised for $7,500? I hope not. While these changes in income are attainable, they are not that easy to achieve.

Remember that the formula is designed to pay the horse out in two and a half years. Should you choose another payout schedule, the values will change. However, determining purchase price by using worst-case scenarios to generate value is much safer than pie-in-the-sky estimates. Anyone who has tried to breed forty mares to a stallion with a $1,000 stud fee will tell you this.

CALCULATION OF YEARLY EXPENSES FOR THE ABOVE EXAMPLE

Boarding fees	$ 7,200
Herd health	400
Farrier	400
Advertising (2 stud fees	2,000
Management fees (2 stud fees)	2,000
Insurance (1 stud fee)	1,000
Total	$13,000

CALCULATION OF YEARLY INCOME

{(number of mares bred – number of barren mares) × stud fee} – 10% for bad debts and another 10% for special deals. Stud fee: $1,000.

	Example 1	**Example 2**	**Example 3**
# of mares bred: income	40/$40,000	40/$40,000	60/$60,000
% live foals: income	50/$20,000	70/$28,000	70/$42,000
10% breeding incentives	–$2,000	–$2,800	–$4,200
10% bad debts	–$2,000	–$2,800	–$4,200
Net income	$16,000	$22,400	$33,600

PURCHASING AN OLDER STALLION

It can be assumed that when a stallion is referred to as an "older horse," the honeymoon is over. This euphemism indicates that the excitement associated with the first three crops has long since gone. In today's fickle stud marketplace, the early years are the most important. Typically, they proceed in the following manner.

Initially, the first year brings in a bevy of mares whose owners are impressed with the prospect's laurels and are full of hope that from this fountain of maleness will spring the next generation's superior offspring. For many horses, their first-year bookings are the largest they will see during their entire careers.

By the second year, with the arrival of the first crop of foals, realism begins to dawn as the foals look very much like an average of the sire and dam instead of showing off the vigor identified with an extremely prepotent sire.

In the third year, shortly after the arrival of the second foal crop, the first crop hits the yearling sales. This further tempers the euphoria originally experienced.

By the time the third crop of foals hits the ground, most young stallions find their bookings noticeably diminished, as skeptical breeders await the real proof of breeding value: How are the two-year-olds doing in training and on the race course? If these first foals have a good or excellent showing, new life will be breathed into the stud career of this great hope. However, if these first runners lack luster, the curtain begins to fall even though physically and reproductively the stallion may just be reaching his most productive years.

Still, for stallions who had fairly active honeymoons, there is still hope as the next couple of decent crops go to the track to try and improve Pop's reputation as a producer of "sure enough" race horses. But, by the fifth year, the mosaic appears almost complete and the verdict is in even though the statistics on those first three crops will not be completed until the eighth year of stud duty.

We can assume that the majority of stallions following this chronological sequence are not superhorses, nor will they be complete washouts. But they will fall into that gray area where it is most difficult to establish breeding value. To help clear this misty view, we now have available a host of numbers by which we can compare one to another. The key to using these statistics is to make sure that we are comparing apples to apples and oranges to oranges. To do this, begin by looking at years at stud and number of foals produced before venturing off into statistics such as

Average per starter
Median per starter
Average per start
Percentage winners from foals

Percentage winners from starters
Percentage two-year-old runners
Percentage stakes winners
Percentage superior runners (foals winning over $25,000)
Average weanling, yearling, and two-year-old prices

It is unfair to compare a horse with ten crops and one hundred foals with a horse who produced one hundred foals in five crops. Obviously the older horse has had more opportunity for more of his foals to reach their maximum performance potential on the track.

Statistics can become even more skewed when they are used to generate additional statistics. The following statistics have been developed to aid in the analysis of Thoroughbreds. (Statistics on other racing breeds can be obtained by contacting individual breed associations.)

Standard Starts Index (SSI)[2]

The Standard Starts Index is a numerical racing index based on average earnings per start for racing performance in North America. SSI compares all horses on the same basis, regardless of their year of birth or sex. The SSI is devised by taking all the foals that were born in a given year and determining the crop's average earnings per start for each year that the crop raced, separately calculating colts and fillies.

These statistics have been compiled for the past fifty years and the following distribution of SSI has been found. For example, if you were interested in a stallion whose SSI was calculated as 2.00, then you would know that only 7.8 percent of all starters would have had a higher index. So it's a way to measure a stallion's race record against the standard.

---

[2] *Developed by Bloodstock Information Services, Inc., Lexington, Kentucky. Information pertaining to Standard Starts Index and Sire Production Index contained herein Copyright Bloodstock Research Information Services, Inc. All Rights Reserved.*

SSI	% Starters Above
32.19	0.1
12.20	0.5
7.61	1.0
4.75	2.0
3.64	3.0
3.02	4.0
2.63	5.0
2.5	5.4
2.00	7.8
1.71	10
1.50	12.4
1.32	15
1.08	20
1.00	20.2
0.91	25
0.77	30
0.69	35
0.57	40
0.49	45
0.42	50
0.35	55
0.29	60
0.23	65
0.16	70
0.10	75
0.05	80
0.01	85

14.6 percent of all starters earned no money and have an SSI of 0.00.

Sire Production Index (SPI)[3]

The Sire Production Index is an average of the Standard Starts Index of all foals by a given sire that have started at least three times in North America. For the SPI to be calculated for any stallion, he must be represented by a minimum of three crops and fifty North American starters.

Average Earning Index (AEI)

The Average Earning Index is a reflection of progeny racing performance compared with the progeny of all other sires for the years raced. Only 2 percent of all North American-based sires have an AEI of 2.00 or higher. This elite group has an average of 10 crops, 32 foals per crop, 7 percent starters, 46 percent winners, 18 percent two-year-old winners, and 10 percent stakes winners. That means 98 percent of all breeding stallions are not in this privileged group!

Comparable Index (CI)

The Comparable Index listed for a sire reflects the racing performance of mares' offsprings produced by stallions *other than the one listed.* Only 30 percent of all sires have an Average Earning Index higher than their mares' CI. Put differently, only 30 percent of all sires produced better race horses out of the same mares bred to other stallions.

We can weigh each of these mathematically produced evaluations to the best of our ability, or perhaps even develop a personalized breeding value index using all of the above statistics combined into one weighted formula. A group of economists, headed by Dr. Robert Lawrence at the University of Louisville's Equine Industry Program, have even

---

[3]*This index as well as the AEI and CI were developed by* The Blood-Horse, *Lexington, Kentucky.*

developed a computer program that takes such values as average yearling price and stud fee and electronically generates an appraisal of a commodity stallion.[4]

All this information increases our chances of making an informed decision in determining a purchase price for an older stallion; a price that allows the new owner a reasonable opportunity to make money with an aged horse. Complicating the economic problem is the fact that the horse is still a biological organism being dealt with by an emotional human being.

Take for example the stallion that has been a mediocre stud. The honeymoon has long since come and gone, his stud fee has been reduced to attract a half-book of mares, and his yearling prices are barely attaining the national mean. Then, along comes an Eclipse Award winner sired from a crop in his teens. The horse's value jumps tenfold at a ripe old age.

Those who own and breed to these older, more quantified gentlemen dream of such a resurgence of economic vigor in the declining years; yet such a reversal is the exception. For the most part, when these data are in, what you see is what you get. Without the dream of a long and illustrious career at stud, the desirability of a mediocre stallion in his declining years decreases.

This is particularly true of stallions in their mid to late teens—for at this juncture, not only is their prowess as producers of race horses in question, but their longevity is also beginning to come under scrutiny. With market demand low, these horses can be a good buy at stud or for purchase. In fact, there may be some remarkable steals for an exceptional horse that never had the opportunity to produce a decent group of offspring before growing old. With a shortened life expectancy at stud, the value of such an individual may have

---

[4]*Additional information about this software may be obtained from Dr. Robert Lawrence, Equine Industry Program, School of Business, University of Louisville, Louisville, Kentucky 40292.*

decreased tremendously, but the opportunity to produce a large number of racing offspring is still there—if one is willing to assume increased risk.

All in all, purchasing the older stallion allows the investor to make a decision based more on reason and logic than on dreams, emotions, and hope. Statistical values determined mathematically leave less room for guesswork when investing in breeding value.

THE STOCK HORSE OR SHOW HORSE–TYPE STALLION

SELECTING A STALLION PROSPECT

Selection of a stock horse or show horse stallion begins with an analysis of his superlative potential in each of the four basic marketing categories of a stud prospect: disposition, conformation, pedigree, and performance. In a survey done a few years ago, several large ranching operations were asked to list, in order, the characteristics for which they selected and bred. Invariably, a stallion's disposition was listed first or second. While I agree with this ranking (and so do most people who are going to ride and train his offspring), disposition is not an easy trait to market as an outstanding reason to breed to a particular horse. Disposition is difficult to quantify; there are no competitions for the best-disposed horse. The owner of the stallion may believe his horse has the best personality in the world, but convincing mare owners of this fact and its economic worth is not an easy task.

Experience has borne this out. I have stood at stud three horses with notoriously bad reputations as mean, ornery cusses: a son of King (AQHA), a son of Sugar Bars (AQHA), and a son of Native Dancer (TB). They were all relatively superior in the other categories, and they all booked mares in sufficient numbers to be profitable stallions—a rare phenomenon.

With the importance of disposition limited to individual interpretation, three areas remain where the stallion prospect can reach the superlative: conformation, pedigree, and performance. It is always possible to breed mares to an absolutely beautifully conformed horse. This does not mean that differences of opinion do not exist on ideal type, but almost all horsemen can agree on those horses that have attained the pinnacle of conformational excellence. And there is a way of competing for the best-conformed breeding animal. In fact, there is a whole industry where conformation is the dominant characteristic in establishing the value of the horse, as well as his stud fee and his appeal to mare owners. For a stallion in this category to gain success with longevity, his offspring must also possess the ideal type of the sire, but in the beginning, the looks and halter record of the horse himself will suffice as the superlative. His wins and grand championships make it relatively easy to establish his superiority over the competition for mares within a specific region or, even, nationally.

Outside of the area of halter horses, conformation still exists as a very marketable trait. Good looks on a barrel horse or a cutting horse will enhance their marketability. Even though mare owners in these categories may be breeding for a performance event as the primary goal, horsemen are generally more amenable to a beautiful animal.

Not only is it more enjoyable to own a well-conformed winner, it makes economic sense. The offspring of a potential racehorse sire that fall short of the mark have more salvage value if they are good looking. While, as the saying goes, there are no ugly horses in the winner's circle, there are lots of good-looking horses that are losers but have not lost their aesthetic value. It should now be clear why one of the characteristics with the broadest economic impact in the selection of the stallion prospect is eye appeal.

A superlative in performance is another matter. Today horses are bred for a number of specialized events and the level of competition in all of them has become very keen. It is an era of specialization. This specialization has led to the selection of specific body types; smaller, slightly built horses for cutting, taller, more powerful horses for jumping.

When seeking the horse with the superlative performance record, we again need to consider the mare population that might be interested in producing for a particular event. For example, the number of sires competing for the title of best cutting horse in Maine is much lower than the number in central Texas. Yet the number of mare owners seeking a cutting horse sire is correspondingly high in Texas. To get a good book of mares in Maine, you would need the top horse in that part of the country, whereas to get a good book of mares in central Texas, you might simply need to have one of the dozen or so good cutting horse sires in the area.

When weighing the importance of performance as one of the superlatives, we should take into consideration the heritability of the trait used in the event. The heritability of speed is estimated at 40 to 60 percent, which is relatively high compared to jumping or pleasure ability. It then becomes evident that the importance and marketability of performance is reflected by a particular event and the mare population seeking to produce offspring for the performance of that event.

The final category is pedigree, since the emphasis on heritability and production performance is oftentimes replaced by names in a pedigree. Pedigree for breeding animals becomes tremendously important in the selection of sires. This is true not only in estimating potential performance, but also in determining the marketability of the offspring (market breeding). To excel in pedigree for the marketing of stud fees, a horse must be by one of the sire lines fashionable at the time. Historically, this will limit the top half of the pedigree to very few horses within a given performance event, while the bottom half is open to a wider range of individual names (although the number of females capable of producing fashionably bred sons is also a rarefied group).

To summarize the conclusions concerning superlatives, it is useful to rank the evaluation categories in order of priority.

1. Conformation
2. Performance
3. Pedigree
4. Disposition

Conformation is listed first and disposition last, but the middle pair, pedigree and performance, are not as easily ordered. A wide range of circumstances makes the two categories interchangeable. If for some reason the horse did not have an opportunity to perform, pedigree becomes more important and, likewise, if the horse performed to the pinnacle of the event, it makes his pedigree fashionable.

Impressive has been dominant as a halter horse sire, yet his sire, Lucky Bars (TB), was not nearly as marketable. Thus Impressive's performance made his appearance in a pedigree more desirable than that of Lucky Bars. There are horses

Doc Bar. Courtesy of the American Quarter Horse Association.

such as King and Doc Bar with fashionable pedigrees for Quarter Horse racing that became founders of dynasties in arena performance, making their pedigrees less significant than their own genetic strength.

While my ranking of importance when selecting a stallion for economic benefit may differ slightly from that of another breeder seeking to produce another desirable type of off-spring, this is the bottom line (as stated earlier): In order to have a reasonable chance of being a commercial success, a stallion must be superlative in at least one of four areas

King. Photo of painting by Oren Mixer Courtesy of the American Quarter Horse Association.

(conformation, pedigree, performance, and disposition) for the population of mares he is likely to attract. If he is excellent in only one of the first three areas, he certainly needs to be at least adequate in the others.

APPRAISING THE VALUE OF AN UNPROVEN STALLION

You just got a phone call from a friend of a friend. "I thought you would be interested in knowing that a certain stallion of the type you like is for sale," the person says. "This stallion has been in competition, but he has peaked out and is now ready for the stud farm. Are you interested?"

Of course, you really had not considered buying a stallion of this quality to breed to your four or five mares, but your curiosity is aroused by this insider information. Before you can stop yourself, you fall into the fatal trap and ask, "Well, how much are you asking for old Stem Winder?"

The answer is a bit much for your pocketbook, but less than you thought they would ask for such a fine animal. So you say, "Let me think on it and I'll get back to you in a couple of days."

No sooner have you put down the phone than you begin to imagine being the proud owner of this champion stud horse—a horse that quite possibly might become a leading sire and make you rich and famous.

Still, you are no greenhorn. You are not going to blindly jump in over your head. Market research—that's the ticket. You begin calling folks with mares to see if they would be interested in breeding to your potential stud. You just can't call everyone you know because you don't want the grapevine to know that this deal is available. So your marketing begins with close friends.

At this point greed appears and chuckles a little at how easily you are taking the bait as you decide that your information about the availability of this animal might be valuable. You can purchase the horse and then sell shares to your friends, making a tidy profit for handling this transaction.

I once knew a man who had the true soul of a horseman and enacted this scenario with the icy coolness of a champion poker player pulling off the classic bluff. After purchasing a stallion with a rubber check for $90,000 on Friday after banking hours, he went home and sold enough shares by Monday morning to cover the check. For putting the deal together, he retained a sizable piece of the horse.

The key for this horseman and for anyone interested in purchasing a stallion revolves around knowing whether or not the quoted price is a good deal. There are several approaches that could be used to pinpoint this value. Which method is best will depend upon the unique circumstances of each individual deal.

Share-Value Appraisal

The purchase price of a stallion selected to sell to one or more partners can be appraised by his potential share value. This concept is best explained by an example. Assume that a limited partnership wants to purchase a stallion and divide him into forty shares. The big question for this group of investors now becomes: What price will these shares bring in the marketplace?

Assuming that the shares will be purchased by mare owners who wish to breed to this particular stallion, it is reasonable to expect that a mare owner should be willing to pay two to five times the stud fee for a lifetime breeding to the horse. Choosing a horse with the potential to book forty to fifty mares at a $500 stud fee should generate a share value of $1,500 and a purchase price of $60,000 (forty shares times $1,500).

The Expenses-Income Method

This approach is most often the method of choice for stallions that do not fit into the category of being a leading money earner or being worth over a half a million dollars in syndication. Evaluating the economics of such stallions takes on a more simplistic estimate of worth:

$$\text{expenses} - \text{income} \times \text{number of years to payout}$$

With an increasing number of stallion hopefuls across the United States standing for a stud fee of $500 or less, let's develop a scenario to calculate his price tag.

The first step is to determine the expenses. Costs of boarding a stallion can range from $5 to $20 a day, depending upon the section of the country and the deal to be made with the stud farm. Assume that the board on this particular stallion will be $10 a day, which will bring his yearly board to $3,600. Add to this basic cost herd health and farrier fees, which are likely to be another $200 per year.

In our expense column we must also include expenses of advertising and management. For horses standing for under $5,000, two breeding fees would be a good estimate of promotional expenses. Similarly, two stud services should be allotted to a management personnel fee and one fee for insurance. This brings the total expenses for a horse standing at $500 into the $6,300-a-year range, or a little over $525 a month.

To begin to estimate the income of this stallion (or any stallion, for that matter), the number of mares that will be bred has to be estimated. Depending on the position of the industry cycle and the geographical location, the statistical average number of mares per stallion hovers around ten. This is an impractical estimate for a profit-making venture, so let's assume that the horse is priced correctly and that he will breed forty mares. This makes the hypothetical income for our $500 per service horse $20,000 a year.

Although this income seems possible, it isn't really realistic. Unfortunately, not all of these hypothetical forty mares will conceive and produce a live foal. While management expertise will play a major role in beating the national average, it would be wise to estimate that 30 percent of the mares will not produce live foals. This will drop our projected income to $14,000.

Also, a decision should be made as to whether this stallion will offer breeding incentives to attract owners of quality mares in order to obtain a bigger and better book. It would be reasonable to expect that this management decision would reduce income by another 10 percent, or in this case, $2,000, causing the resulting income to drop to $12,000.

And, a value should be factored into the formula to account for the bad debts that are so prevalent in the breeding business. Personal experience estimates that a 10 percent loss of income is reasonable to expect.

Let's look again at the calculation for income:

{(number of mares bred – number of barren mares) × stud fee} – 10% for bad debts and another 10% for special deals

This brings the income for our example to $10,000. Add these figures to the original formula and we get the following information:

$10,000 – $6,300 = $3,700 × number of years to payout

If we assume that the net profit of the horse (in this case $3,700 a year) will be applied to the purchase price, how many years can you afford to pay for the horse? While the IRS allows a young breeding animal to be depreciated over seven years, most breeders would find it a financial disaster to stand a horse for seven years without making any profit.

In my experience, a reasonable amount of time for a horse to pay himself out is two and a half years. Using two and a half as the number of years to payout, we now have sufficient information to complete the computation and estimate the value of this stallion based upon the original postulation:

$10,000 – $6,300 = $3,700 × 2.5 = $9,250

As I said before, an example such as this confirms the validity of the expression "It's hard to make money with cheap horses." Booking forty mares to a $500 stallion is not easy and the quality of management will make the difference. This can easily be seen by changing a few values.

Suppose this horse produced 90 percent live foals instead of 70 percent. His income would become $14,000 instead of $10,000 and his estimated value would become $19,250.

Or suppose that he booked sixty mares with a 70 percent production rate; the income would be $15,000 and he'd be worth $21,750.

Would you now be willing to spend $19,250 or $21,750 for a horse originally appraised at $9,250? I hope not. While these changes in income are attainable, they are not that easy to achieve.

Remember that the formula is designed to pay the horse out in two and a half years. Should you choose another pay-out schedule, the value will change. However, determining purchase price using worst-case scenarios to generate value is much safer than wishful thinking, as anyone will tell you who has tried to breed forty mares to a stallion with a $500 stud fee.

CALCULATION OF YEARLY EXPENSES FOR THE ABOVE EXAMPLE

Boarding fees	$ 3,600
Herd health	75
Farrier	125
Advertising (2 stud fees)	1,000
Management fees (2 stud fees)	1,000
Insurance (1 stud fee)	500
Total	$ 6,300

CALCULATION OF YEARLY INCOME

{(number of mares bred – number of barren mares) × stud fee} – 10% for bad debts and another 10% for special deals. Stud fee: $500.

	Example 1	Example 2	Example 3
# of mares bred: income	40/$20,000	40/$20,000	60/$30,000
% live foals: income	70/$14,000	90/$18,000	70/$21,000
10% breeding incentives	–$2,000	–$2,000	–$3,000
10% bad debts	–$2,000	–$2,000	–$3,000
Net income	$10,000	$14,000	$15,000

APPENDIX

SAMPLE CONTRACTS

SAMPLE CONTRACTS

The following contracts are samples of different types of legal instruments that are commonly used in business transactions involving stallions. These contracts and my comments are meant only to familiarize you with typical agreements and to point out areas of special interest to the horseman. For more specific information, consult an attorney familiar with equine law in your state.

THE BILL OF SALE
FOR BREEDING STALLION

When buying a stallion, the standard consideration for a bill of sale should be spelled out in the simplest way. In addition to what individual states require for the bill of sale, unique specifications may exist in the sales agreement for the particular stallion.

The guarantee of fertility is one of the most common additions to an agreement. A specific definition of exactly what constitutes fertility (see Chapter 12) should be given. The sales agreement should also specify the measures to be taken should the horse fail to pass his fertility test.

Another matter of consideration is how and when money is expected to change hands, the exact time when the purchaser assumes the liability for the animal, and when the new owner assumes all risk of injury or death.

Having the foregoing information clearly set forth in a contract may help prevent problems in the buyer-seller relationship.

Sample
BILL OF SALE
for
BREEDING STALLION

Stallion's name: Years Away
Date of contract: October 3, 1986

For and in consideration of Five Thousand Five Hundred Dollars ($5,500.00) I, James P. McCall of The Old Place, Mount Holly, Arkansas, 71758, hereby sell, assign, and transfer to Lynda Blair of Jimani Farm, Route 4, Box 140 B, Ruston, Louisiana, 71270, ninety-five percent (95%) of my entire right, title, and interest in and to the 1979 bay Thoroughbred stallion Years Away, by Riva Ridge and out of Contrarian by Dewan.

The seller guarantees to be the sole owner of the above horse and that the above horse is free and clear of all encumbrances.

The seller guarantees that the stallion is free of severe vices such as weaving and cribbing and is breeding sound.

By retaining five percent (5%) of the stallion Years Away, the seller will share up to his pro-rated portion in the proceeds, should Years Away be sold. In addition, the seller will be responsible for five percent (5%) of the expenses of the stallion. Five percent of the stallion will also entitle the seller to two (2) breedings per year.

Upon the mutual signing of this contract, the buyer agrees to forward a good-faith check for Two Thousand Dollars ($2,000.00). The seller agrees to furnish at that time, for the aforestated horse, a current negative Coggins and all necessary health papers for shipping. Upon notification by the seller that these items have been received, the buyer will arrange for the transport of the stallion to Louisiana.

Upon Years Away's arrival in Louisiana, the buyer will send the seller a check for Three Thousand Five Hundred Dollars ($3,500.00). Upon its receipt the seller will send by registered mail the foal certificate for Years Away.

Seller_____ Date_____

Buyer_____ Date_____

Witness_____ Date_____

STALLION LEASE AGREEMENT

While stallion lease contracts can have a wide range of terms for payment of the lease, generally all contracts cover certain common matters. Those entering into the agreement should determine, among other things:

- which party will be responsible for the cost of the upkeep, such as the board.
- if insurance is to be kept, who is to pay the premium.
- whether the lessee or lessor will be responsible for any medical bills.
- who will assume liability for the animal.
- which party will plan and/or pay for the advertising campaign.

It should also be pointed out that, in many cases, it is common for the person standing the stallion to either retain a percentage of the horse or be allowed some breeding privileges without compensation to the lessor.

Sample
STALLION LEASE AGREEMENT

This lease is by and between James and Lynda McCall, hereinafter referred to as LESSOR, and Jimani Farm of Ruston, Louisiana, hereinafter referred to as LESSEE.

LESSOR does hereby lease to LESSEE and LESSEE does hereby lease from LESSOR, for a period beginning the first day of October 1993, and ending the last day of September 1994, the following stallion:

Name: Grand Slam Dan

Age: 12

Registration number:_____

LESSOR warrants that he has good and clear title to aforesaid stallion free of any liens.

The parties hereto hereby agree that title and ownership of aforesaid stallion shall be and remain in the name of the LESSOR.

LESSEE hereby agrees to maintain aforesaid stallion in good condition, with proper care, handling and protection according to the rules of good animal husbandry and reasonable standards and practices of the horse breeding industry and shall assume financial responsibility for such maintenance at no cost to the LESSOR.

LESSEE hereby agrees that LESSOR or his authorized agent may go upon LESSEE'S premises at any time and inspect aforesaid stallion and determine if same is being properly cared for and in good health.

LESSEE hereby agrees that all of the revenues generated as stud fees by aforesaid stallion from all mares booked and bred, less ten percent (10%), is and shall be the property of the LESSOR. This booking fee and revenues generated by mares booked to aforesaid stallion, such as from board and veterinary care, are and shall be the property of the LESSEE. It is further agreed that the LESSEE may breed up to five mares to the aforesaid stallion without payment of stud fee.

LESSOR and LESSEE shall annually agree on the stud fee.

LESSOR hereby agrees to pay for any and all advertising of aforesaid stallion; said advertising shall be agreed upon by both the LESSOR and LESSEE and shall be not less than Five Thousand Dollars ($5,000.00) per annum.

LESSOR hereby agrees that LESSEE is not required to provide death or injury insurance on aforesaid stallion.

LESSOR hereby agrees to pay the cost of any and all major medical expenses accrued by the stallion, this cost to include any and all medical expenses that exceed Fifty Dollars ($50.00), during the period described by the terms of the contract.

LESSEE hereby agrees to hold LESSOR harmless from any claim resulting from damage or injury caused by aforesaid stallion.

Should aforesaid stallion become missing, lost, estrayed, seriously injured, sick, or dead, LESSEE hereby agrees to notify LESSOR immediately.

LESSOR hereby agrees to hold LESSEE harmless from any claim resulting from accidental injury or death of aforesaid stallion.

It is further agreed that the LESSOR is the authorized agent in signing breeders' certificates and stallion breeding reports of aforesaid stallion.

This agreement is for one year. If either party determines to terminate the agreement after one year, notification of such action will and shall be made ninety (90) days prior. If no such action is taken, the agreement will continue on a year-to-year basis.

This lease and all covenants and agreements herein contained shall accrue to and be binding upon the parties hereto, their heirs, successors, administrators, and executors.

LESSOR	LESSEE
Name	Name
_____	_____
Title	Title
_____	_____
Address	Address
_____	_____
_____	_____
Telephone number	Telephone number
_____	_____

STALLION SERVICE CONTRACT

There are ten general categories that should be considered in adopting a stallion breeding contract for a particular management:

1. All of the people involved should be named.
2. Horses involved should be correctly identified.
3. Stud service compensation and guarantees should be completely spelled out.
   a. Live foal option requires that a foal to be considered alive must stand and nurse. If a live foal does not result from the breeding, the contract should describe compensations due the mare owner. For example, if the fee was prepaid, the money may be returned, or the mare owner may receive return privileges without the option of a cash refund. Regardless, the exact date of payment of fees should be described in the contract (e.g., due at time of service, due when the foal stands and nurses, one-half due on signing of contract, the other half due September 1, etc.).
   b. Nonguaranteed breedings indicate that the fee is to be paid upon booking and there will be no refund if the mare is covered by the stallion.
4. Times at which board fees and other expenses are due.
5. Description of boarding costs such as nonrefundable breeding shed costs (chute fee) or booking fees.
6. Liability disclaimers.
7. What constitutes contract termination.
8. Mare refusal clause.
9. Type and quality of care for mare and/or foal.
10. An abandonment clause: What happens to the mare if the bills are not paid and the mare is not picked up.

Sample
STALLION SERVICE CONTRACT

This agreement is by and between James P. McCall, hereinafter referred to as Stallion Owner, and Saundra Lewis, hereinafter referred to as Mare Owner. Stallion owner hereby agrees to breed the Thoroughbred stallion Years Away to the mare Twice Happy, Breed TB, Registration No._____, Sire: Twice Worthy. Dam: Born Happy, which is owned or leased by Mare Owner during the breeding season of 1994.

Mare Owner hereby agrees to pay the following fees for service pursuant to this agreement.

1. A nonrefundable boarding fee of Fifteen Dollars ($15.00) per day shall be charged to Mare Owner by Stallion Owner for each day aforestated mare (and foal) are at the Farm, excluding dates of arrival and departure. Said boarding fee shall be paid before aforestated mare (and foal) leave the Farm.

2. Nonrefundable veterinary and farrier fees shall be charged to Mare Owner by Stallion Owner for appropriate services for aforestated mare (and foal). Said fees shall be paid before aforesaid mare (and foal) leave the Farm.

3. A booking fee of Two Hundred Dollars ($200.00) shall be due upon signing of this contract and is nonrefundable.

4. A stud fee of One Thousand Five Hundred Dollars ($1,500.00) shall be charged for service. This fee is due and payable when the aforestated mare has a live foal that stands and nurses. If the aforestated mare fails to have a live foal that stands and nurses, or has twins, the mare may be returned to the service of the stallion without charge of additional stud fees, provided a licensed veterinarian attests to such occurrence in writing and notification is to the Stallion Owner within one week of such occurrence.

Mare Owner hereby agrees that Stallion Owner reserves the right to refuse service to any mare which appears to be diseased or which is unfit for breeding, in the opinion of the Stallion Owner.

It is agreed by the parties hereto that if the aforestated stallion should die, be sold or become unfit for breeding before serving aforestated mare, or if aforestated mare should die, be sold or become unfit for breeding before served, this agreement is and shall be null and void, except for provisions in paragraphs 1 and 2.

Mare Owner hereby agrees to hold Stallion Owner harmless from any claim resulting from accidental injury or death of aforestated mare (and foal).

Should aforestated mare (and foal) become missing, lost, estrayed, seriously injured, sick, or dead, Stallion Owner hereby agrees to notify Mare Owner or his authorized agent immediately by telephone and subsequently by mail.

Stallion Owner hereby agrees to hold Mare Owner harmless from any claim resulting from damage or injury by aforestated mare (and foal).

Stallion Owner hereby agrees to maintain aforestated mare (and foal) in good condition pursuant to this agreement, with proper care, handling, and protection, according to the rules of good animal husbandry and reasonable standards and practices of the horse industry.

Stallion Owner hereby agrees that Mare Owner may go upon Stallion Owner's premises at any time and inspect aforestated mare (and foal) and determine if same is in good health and properly cared for.

IF ARRANGEMENTS ARE NOT MADE TO PICK UP MARE (AND FOAL) AND ALL INCURRED EXPENSES PAID BY SEPTEMBER 1, 1994, THE AFORESTATED MARE (AND FOAL) WILL BE CONSIDERED ABANDONED AND WILL BE DISPOSED OF AT PUBLIC AUCTION.

This agreement and all covenants and agreements herein contained shall accrue to and be binding upon the parties hereto, their heirs, successors, administrators, and executors. Upon material breach of this agreement by one party, the other party may terminate same.

No modification of this agreement shall be binding unless in writing and executed by the parties hereto.

This agreement is subject to the laws of the State of Arkansas, executed on January 1, 1994.

Mare owner or agent   Stallion owner or agent

_____   _____

Address   Address

_____   _____

_____   _____

Telephone number   Telephone number

_____   _____

STALLION SYNDICATION AGREEMENT

One of the primary considerations in preparing a stallion syndication is to ensure that those participating in the limited partnership purchase are aware of the standard practices in the industry. For a group of experienced and informed individuals with previous experience in owning stallion shares, there might not be a need for a contract quite as explicit as the following example, which was prepared for a group of investors having little experience with this type of combined ownership.

It is also important to note that active participation in the management of the syndicate produces different tax benefits than are available to the nonparticipating shareholder.

Readers should also be aware that stallion syndication is not the only option for group participation in stallion ownership. Stallions can be incorporated or held by groups of investors bound together by various other legal contracts. To further explore these options, seek the advice of a contract attorney or CPA to determine the advantages and disadvantages of each particular situation.

Sample
STALLION SYNDICATION AGREEMENT

THIS AGREEMENT is entered into as of this second day of June in the year Two Thousand, by and between James P. McCall of Union County in the State of Arkansas, hereinafter called "Owner," and the signers of the counterparts hereof, hereinafter called "Shareholders";

WHEREAS the said James P. McCall is the record owner of the Thoroughbred stallion Jimani Ruler, and has agreed to sell the stallion to a syndicate to be formed hereunder; and

WHEREAS the parties desire to join together for the purpose of acquiring, owning, and holding Jimani Ruler for breeding purposes to their mutual advantage;

NOW, THEREFORE, THIS AGREEMENT WITNESSETH, That for and in consideration of the mutual promises thereinafter set forth, the parties mutually agree as follows:

1. PURPOSE
The parties hereby form this syndicate for the purpose of buying, owning, and holding for breeding purposes the Thoroughbred stallion Jimani Ruler. The name of the organization hereby formed is "Jimani Ruler Syndicate," hereinafter referred to as "syndicate."

2. OWNERSHIP AND PRICE
The parties hereto mutually agree to and hereby do form a syndicate to purchase the Thoroughbred stallion Jimani Ruler at and for the price of TWO HUNDRED THOUSAND DOLLARS ($200,000.00), and the Owner hereby agrees that the ownership of Jimani Ruler shall vest in twenty-five (25) shares of this syndicate; at the time twenty-one (21) or more shares, exclusive of the Owner's shares, shall have been subscribed for by shareholders signing this Agreement and paying for the said shares as hereinafter described. The parties agree that Owner may transfer legal ownership of the stallion to the syndicate of record or may hold record title in himself, but that, in any event, upon completion of the payment terms and conditions of this Agreement, the shareholders shall be the beneficial and equitable owners of Jimani Ruler, and if Owner retains record title in the horse he will nevertheless hold title to Jimani Ruler for and on behalf of the shareholders. Owner may retain up to five (5) shares in the syndicate for himself; (five nonnegotiable breedings will be transferred to the breeding farm that stands Jimani Ruler in exchange for the expenses of standing the stallion, including but not

limited to feed, board, veterinary services, etc.) and the Owner shall be entitled to all the rights and privileges of a shareholder herein and shall share in all the obligations, pro rata of a shareholder.

3. SHARE PAYMENT—TERMS AND CONDITIONS

Each shareholder shall pay to Owner Eight Thousand Dollars ($8,000.00) for each share in cash in one installment at the time of executing this Agreement. There shall be twenty-five (25) shares; each share shall be on an equal basis with all other shares and shall be indivisible. No ownership interest in a share shall be transferred to another person without the consent in writing of the owners of a majority of the shares, and upon such consent the assignee shall thence become a shareholder with all the rights and obligations thereof. No individual shareholder shall have any right or power to require or cause, except by exercising his right to vote his share as herein provided, the sale or transfer of the stallion or to grant or give any security interest in the stallion, each share being entitled only to the rights and privileges stated herein to the end that no individual shareholder or his assignees or transferees by operation of law shall have the right to require or cause the stallion to be sold by court order, levy, or judgment or otherwise. The shareholders, or any one of them, by their majority consent, shall have the right to purchase at fair market value any share acquired by an assignee or transferee by operation of law, such as a personal representative, trustee in bankruptcy, receiver, guardian, or similar legal representative, or of any judgment owner or lien holder. Or, if a majority of the owners of the issued shares so agree in writing, the said assignee or transferee may be permitted to retain the said share with all the rights and all of the obligations of a shareholder herein. Notwithstanding the provisions of this paragraph, the interest of any shareholder may be transferred or disposed of by will or intestacy to or for the benefit of the deceased shareholder's immediate family or may be transferred during his life by gift or inter vivos trust to or for the benefit of his immediate family. For the purposes of this paragraph, "immediate family" is defined as the spouse, children, or grandchildren of the shareholder.

Each shareholder shall be entitled to one breeding of a mare to Jimani Ruler during each annual breeding season, and the right to each annual breeding may be assigned by the shareholders without the consent of the other shareholders. Should a shareholder fail to use or sell or transfer his annual breeding privilege by April 1st of the breeding season, the said privilege will revert to a pool of unused shareholder breedings. Each shareholder shall notify the syndicate manager by April 1st of each breeding season if he intends to have his own mare be bred by Jimani Ruler or if and to whom he has sold his breeding privileges or that his breeding privilege is to be placed in said pool. If the privilege is placed in the said pool, the syndicate manager will proceed to sell the same, applying the fees therefore equally to the account of those shareholders who have placed their annual breedings in the said pool. Should it be determined, for any reason, that Jimani Ruler cannot breed thirty (30) mares during each annual breeding season, then shareholder breeding privileges shall be chosen by lot or drawing at a meeting, after due notice of not less than seven (7) days, of a majority of the owners of the shares who shall then determine the procedure of the lot or drawing. All fees for annual breeding services above thirty (30) shall be deemed a syndicate income and shall be disposed of as hereinafter provided.

4. INCOME AND EXPENSES

All income from over thirty (30) breeding fees shall be used first, to pay the expenses of insurance, advertising, etc., and, second, any surplus therefore shall be distributed annually during the last week in December to the shareholders in proportion to which their respective shares bear to the whole number of shares. Should syndicate expenses exceed income at any time, the shareholders hereby agree that periodically, but not more frequently than monthly, such outstanding expenses shall by the manager be divided by the number of shares and an account thereof sent to each shareholder. Shareholders agree to pay their pro rata share of the said expenses within thirty (30) days of receiving the said statement of account, and each shareholder further expressly agrees that if said expenses are not paid within thirty days, the syndicate may charge and receive, and the shareholder shall pay, an extra charge of one and one-half percent (1.5%) each month of the balance of the amount remaining unpaid, or an equivalent annual interest rate of eighteen percent (18%) on the unpaid balance.

5. SYNDICATE MANAGER

The Owner shall assume the position of syndicate manager; the syndicate manager may be paid such sum or sums by the syndicate for his service as the majority of the shareholders agree. The shareholders may choose to remove the Owner from the position of syndicate manager by two-thirds of the owners of the shares of the syndicate. The syndicate manager shall have the sole direction, management, and entire conduct of the syndicate in accordance with the Agreement. Jimani Ruler shall stand at such places as the manager may determine, under the sole and personal management, care, and supervision of the said syndicate manager, who shall select the veterinarian, establish the annual stud fee, and the conditions of service, and advertise the stallion. The manager may, for and on account of the syndicate, select depositories, employ agents, counsel, clerks, accountants, and other necessary assistance, and may acquire and use all and any syndicate funds and pay any and all expenses for the purposes of the syndicate and in connection therewith. The manager shall keep Jimani Ruler insured against such risks as may be economically insurable for the benefit of the shareholders as their interest may appear.

The manager may compromise, settle, or arbitrate any claims in favor of or against the syndicate, but he may not without the written consent of three-fourths of the owners of the shares, sell or pledge or permit anyone to acquire any security interest in the stallion. The manager shall not have the power to enter into any obligation which may involve the members of the syndicate in any liability in addition to that expressly imposed by the terms of this Agreement. If, however, at any future time it is considered in the interest of the syndicate to incur any extraordinary or capital expenditures, it shall first be submitted to a special meeting of the syndicate called for that purpose, of which not less than seven (7) days notice in writing shall be given and ratified and adopted by at least three-fourths of the owners of the shares, whereupon it shall become binding upon all members of the syndicate.

6. PRINCIPAL OFFICE
The principal office of the syndicate shall be established by the manager. The manager shall cause to be kept at the principal office sufficient and accurate records as may be necessary to reflect the true financial and contractual condition of the syndicate at any time; he shall also keep a current register of the members of the syndicate, and each shareholder shall keep the manager advised of his current address and telephone number. Any and all records of the syndicate shall be open to inspection to any syndicate member at any time during normal business hours. The funds of the syndicate shall be deposited in such bank account or accounts as may be required and the manager shall arrange for the appropriate conduct of all accounts. All withdrawals shall be signed by the manager. Annually, in the last week of December of each year, the syndicate manager shall mail to each shareholder a report of the syndicate's activities for the year then ending, including a statement of income and expenses, a list of breedings, and any other pertinent information.

7. TERMINATION OF THE SYNDICATE
The syndicate shall be dissolved and terminated upon the occurrence of any of the following events:

a) The death or any disability of Jimani Ruler rendering him unfit for breeding purposes.

b) Whenever the owners of three-fourths of the shares of the syndicate shall determine in writing that Jimani Ruler shall be sold.

c) Whenever the owners of three-fourths of the shares of the syndicate shall determine in writing that the syndicate shall be terminated.

Upon the termination of the syndicate the manager shall proceed to sell all of the syndicate's assets upon the terms and conditions agreed to by the majority of the owners of the shares, and, after deducting all outstanding syndicate expenses, including the expenses of the sale, the manager shall distribute the net proceeds thereof to the then shareholders of record, pro rata.

8. MEETINGS AND VOTING
At any meeting of the shareholders or upon any question to be decided, each share shall be entitled to one vote in person or by proxy; voting may be by voice or by ballot as a plurality of the members attending shall decide. Adequate notice in writing of at least seven (7) days of the time and place of and the precise questions to be decided at the meeting shall be given to each shareholder at his last known address by ordinary mail. The manager may call a meeting by giving such notice; the owners of one-fourth of the shares by signing and mailing such notice may call a meeting. The manager shall preside at any meeting of the shareholders; in his absence, the shareholders shall elect a temporary chairman to preside at the meeting.

Any action that may be taken at a meeting of the shareholders may be taken without a meeting if a consent in writing, setting forth the action so taken or contemplated shall be signed by all of the shareholders entitled to vote with respect to the subject matter thereof.

Where "majority" or "three-fourths" or "unanimous" consent is used in this Agreement, it shall mean the consent, by recorded vote at a meeting or in writing signed by the shareholder of a majority or thirteen (13), or three-fourths or nineteen (19) of all twenty-five (25) as appropriate, of the twenty-five shares.

9. GENERAL
Each shareholder, by signing this Agreement, warrants to all other parties that he has independently informed himself of the bloodlines and breeding potential of Jimani Ruler, that he is aware of the speculative nature of this venture and is able to afford any losses that may be incurred, that he is entering this Agreement for investment purposes, that he appreciates and understands that the stallion may become impotent or that his get may fail to fulfill his potential or he

may be injured or that he may die at any time after this Agreement is executed, and that in such event or events, the shareholder may lose his entire investment plus any pro rata share of expenses and costs that the syndicate may have incurred. Nothing contained in this Agreement or otherwise shall constitute the syndicate's subscribers' partners with the syndicate manager or with the Owner or with one another, or render them liable to contribute more than their ratable amounts as specified herein or entitle them to any participation in results of profits or losses of the syndicate other than as specified in this Agreement.

This Agreement may be executed in several counterparts, and when executed by all the shareholders and accepted by the Owner, the several parts shall constitute the Agreement between all of the parties as if all signatures were appended to one original instrument. Where appropriate the singulars shall be read as the plural, the plural as the singular, the masculine as the neuter or feminine, the neuter as the masculine or feminine, or the feminine as the masculine or the neuter. All paragraph titles or captions contained in this instrument are for convenience only and shall not be deemed part of this Agreement. This Agreement constitutes the entire agreement among the parties and may not be modified unless by a line instrument in writing. This Agreement shall be binding upon and, subject to the limitations expressly hereinbefore set forth, shall inure to the benefit of all the parties, their personal representatives and assigns.

IN WITNESS WHEREOF, the parties have hereunto set their signatures, seals, addresses and number of shares to which they have subscribed:

Number of Shares   Shareholder

_____4_____   _____

                         James P. McCall, Owner
                         McCall Road
                         Mount Holly, AR 71758

POSTSCRIPT

RAISING A YOUNG STALLION FOR PERSONAL USE

AN ILLUSION WRAPPED IN AN ENIGMA
SHROUDED IN MYSTERY

Owning a stallion for pleasure is one of the great enigmas of the horse/human relationship. Many horse owners and potential horse owners have a dream of establishing a meaningful and beneficial relationship with a mature male horse.

I have often wondered about this. No doubt Hollywood has contributed to this dream with films like *Smokey* and *The Black Stallion*. Fictional works such as the Black Beauty books perpetuate the vision.

At one time, I believed this desire for such a relationship to be more ingrained in the male psyche than the female. Perhaps the macho image of riding a stallion off into the sunset had appeal. Or perhaps it was the image of riding a fierce charger into battle, striking envy and awe into the hearts of the enemy. Today, however, I think differently because it seems that more women than men want to own and ride stallions.

I can't come up with a reasonable sociocultural theory for this trend. Could it be that the rationale is based on either logic or economics? For example, I know of many owners who have purchased a male horse as a weanling in the hope that he would grow up to be a particularly outstanding individual. For this reason, it appears to be unthinkable to them to prevent this colt from being able to pass on his genetic material, especially if someone is willing to pay for it.

Also on the side of logic are the biological benefits of leaving the production of male sex hormones intact. Stud horses are known to have more muscle mass, which should give them strength and stamina. It is also said that stallions have more strength of character, which when controlled makes them fierce competitors. Unfortunately, uncontrolled or untrained, this same attitude becomes a curse in the day-to-day experience of living with a stallion.

These feeble attempts do not explain the obvious. Today, more than at any other time in the history of the recreational horse, male horses are being left whole even though they are not being used for breeding. The current view seems to be: Don't fix a colt unless he is broken. In other words, if there is no clear reason to castrate, then one should not.

This practice results in many trainers having at least a couple of colts in their barn at any given time that have not been gelded. Commonly, the owner is a woman who either raised or bought a sexually immature colt that she nurtured and trained until he reached eighteen to twenty-four months.

Somewhere during this time, the gentle and pliable little creature began to take a few too many liberties. Its oral fixation started as a lipping curiosity but progressed to nips that included putting five hundred pounds of pressure per square inch onto the body.

This same gentle and quiet youngster may have been ridden by his owner until he became more interested in communicating with other horses—a fact that made him less interested in taking direction from his owner. The colt became less controllable, and a professional trainer was sought out to alleviate the problem.

A trainer is somewhat like a veterinarian. A vet mainly sees sick and diseased horses, since healthy horses are less in need of his services. Likewise, a trainer deals with a biased sample of the equine population in that the horses sent to him are in need of some specialized training. If a horse performs desirably for the owner, a trainer's services are in less demand. This explains why hundreds of presumptuous colts have come into my life. My experience has taught me that the application of some general rules will help keep a stud horse's relationship with humans on a congenial and gentlemanly basis.

First, as the young adolescent male horse starts to have his maleness come to the forefront, he will quite naturally go into a stage of oral fixation and want to taste or bite his companions and handlers. A colt should never be allowed to develop a habit of pinching human flesh. The first time a yearling colt starts to put his lips against a person's skin or clothes, it is imperative to abruptly interrupt this behavior. Reprimand him with a harsh word and a rap on the muzzle. Left unchecked, there is a 90 to 100 percent chance that this innocent nibbling and nuzzling will turn into nipping and biting. While this behavior may seem only mildly irritating when delivered by a youngster, an adult male is capable of inflicting severe pain and injury.

"Not my horse," you are likely to reply. "He likes me. He would never do anything to hurt me." During the past thirty years that I have been standing breeding stallions and breaking and training young colts, I have never known a stud colt, no matter how gentle, no matter how kind, that didn't make at least one attempt to bite someone. Granted, sometimes it seems like they really didn't want to try, but their genetic program made them do it. But unless they are convinced that they should never again aim an oral attack at a human, it is going to happen again. The sooner you convince them, the easier life is going to be for both the horse and his human companions.

Second, expect stud colts to challenge your right to be their masters. Biogenetically, their instincts tell them that as they mature they must keep challenging authority figures. The goal of this genetic program is directed toward dominating a herd of mares and achieving and maintaining the role of herd sire through combative confrontations (during which biting plays a major role).

Being aggressive, therefore, is part and parcel of a stallion's behavior. Fortunately, in the beginning, nature did not create all stallions with the same level of aggression. The intervention of man in establishing the criteria necessary for the perpetuation of the species has also played a role in the reduction of the level of aggression in stallions. However, the fact remains that nature has had millions of years to establish its genetic codes and man has been intervening for, at the most, seven thousand years—a mere heartbeat in evolutionary time.

This means that it is highly likely that a stallion is going to want to push on your line of control and authority from a very early age. This behavior is going to continue until the horse is totally convinced that he is not going to assume the rank of the dominant animal or until he becomes afraid to keep trying.

As an owner, trainer, or handler of stallions, you must have a very clear image of the behavior you demand from the horse. Once you draw the line, it cannot be changed. The horse is going to push or test the line and you must be prepared to push him back.

The amount of aggression or force that is needed depends on the disposition of the particular male horse in question. The proper response for an aggressive five-year-old stud would probably traumatize an eighteen-month-old colt. As with all horse training, feel and finesse are the primary tools that will help bring the young stallion under control.

Third is the problem of socializing the animal. It is normal and natural for the male horse to want to meet and greet other horses, especially females. The desire for this type of companionship, which occurs 365 days a year in stallions, is opposite that of mares, who only seek male company during the receptive period of their estrus cycle. In other words, stallions are normally in heat all the time. This makes them very social animals with strong opinions about the makeup of the group they wish to be with.

Unfortunately, many stallion owners are not prepared to deal with these socialization needs. The standard practice is to isolate the male in a stall and/or paddock—a situation that makes it impossible for the quarantined individual to play out his hormonally determined role.

This management routine is even more difficult on the young stallion. Programmed to learn the social graces of the equine herd through play fighting and interaction with adults, the isolated youngster is overloaded with energy. Ignorant of proper behavior, it falls to his human owner/trainer to teach the colt how to act in the presence of other horses—a task of immense complexity that is often compounded by the naiveté of the novice owner.

One solution to this problem is to contrive circumstances where the colt can be turned out with other more mature individuals, so they can have a hand in his education. An older gelding may simplify your existence. Even being turned out with young males of the same age will allow the stud colt to satisfy his need for companionship. All of these solutions will make it easier to handle the horse when the time comes for horse/human contact.

And, if you own a stallion, that time is going to come. In order for this horse to become a useful, pleasurable mount, someone eventually must teach this horse exactly what is acceptable behavior while in the company of humans and horses of all sexual leanings, such as mares, mares in heat, geldings, and other stallions. To do this you must dominate the actions of the colt. He must know that regardless of the composition of the equine group that he finds himself in, *you* are the herd boss and *you* control his behavior, no matter what.

Given that you can establish and maintain the dominant role in this horse/human relationship, many stallions can become enjoyable partners with a certain uniqueness that only an intact male can provide. Achieving this goal often depends on whether the stallion has a temperament that allows him to be controlled and dominated with enough ease that his maleness is more of an asset than a fault. When this doesn't happen, the joy of owning a stallion becomes an illusion.

BIBLIOGRAPHY

Amann, R. P., and V. K. Ganjam. "Effects of Hemicastration or HCG Treatment on Steroids in Testicular Vein and Jugular Vein Blood of Stallions." *Int. J. Androl.* 3 (1981):132–39.

Amann, R. P., D. L. Thompson Jr., E. L. Squires, and B. W. Pickett. "Effects of Age and Frequency of Ejaculation on Sperm Production and Extra Gonadal Reserves in Stallions." *Journal of Reproductive Fertility Suppl.* 27 (1979):1–6.

Asbury, A. C., and J. P. Hughes. "Use of the Artificial Vagina for Equine Semen Collection." *Journal of the American Veterinary Medical Association* 144 (1964):879.

Bedrak, E., and L. T. Samuels. "Steroid Biosynthesis by the Equine Testis." *Endocrinology* 85 (1969):1,186.

Berndtson, W. E., B. W. Pickett, and T. M. Nett. "Reproductive Physiology of the Stallion, IV: Seasonal Changes in the Testosterone Concentrations of Peripheral Plasma." *Journal of Reproductive Fertility* 39 (1974):115–18.

Berndtson, W. E., J. H. Hoyer, E. L. Squires, and B. W. Pickett. "Influence of Exogenous Testosterone on Sperm Production, Seminal Quality and Libido of Stallions." *Journal of Reproductive Fertility* Suppl. 27 (1979):19.

Berndtson, W. E., E. L. Squires, and D. L. Thompson Jr. "Spermatogenesis, Testicular Composition and the Concentration of Testosterone in the Equine Testis as Influenced by Season." *Theriogenology* 20 (1983):449–57.

Bielanski, W. "Reproduction in Horses." *Stallions, Instytut Bulletin* (Kracow, Poland: Instytut Zootechniki) 1 (1960):116.

Bielanski, W., and F. Kaczmarski. "Morphology of Spermatozoa in Semen From Stallions of Normal Fertility." *Journal of Reproductive Fertility* Suppl. 27 (1979):39.

Blue, B. J. "Effects of Pulsatile or Continuous Administration of GnRH on Reproductive Functions of Stallions." Master's thesis, Colorado State University, 1990.

Blue, B. J., et al. "Effects of Pulsatile or Continuous Administration of GnRH on Reproductive Functions of Stallions." *Journal of Reproductive Fertility* Suppl. 44 (1991):145–54.

Burns, P. J., et al. "Effects of Season, Age and Increased Photoperiod on Reproductive Hormone Concentrations and Testicular Diameters in Thoroughbred Stallions." *Reproductive Endocrinology* 4, no. 5:202–8.

Burns, P. J., and R. H. Douglas. "Effects of a Single Injection of GnRH or Equine Pituitary Extract on Plasma LH and FSH Concentrations in Stallions." *Journal of Reproductive Endocrinology* 4, no. 6:281.

———."Reproductive Hormone Concentrations in Stallions With Breeding Problems: Case Studies." *Journal of Reproductive Endocrinology* 8, no. 1:40.

Dickson, V., J. Schumacher, T. L. Blanchard, and L. Johnson. "Diseases and Management of Breeding Stallions." *American Veterinary Publications.* (1991)

Douglas-Hamilton, D. H., R. Osol, G. Osol, D. Driscoll, and H. Noble. "A Field Study of the Fertility of Transported Equine Semen." *Theriogenology* 22 (1984):291.

Frerichs, W. M. "Effects of Imidocarb Dipropionate and Hemicastration on Spermatogenesis in Pony Stallions." *American Journal of Veterinary Research.* 38 (1981):39–41.

Ganjam, V. K. "Episodic Nature of the ^4-ene and ^5-ene Steroidogenic Pathways and Their Relationship to the Adreno-gonadal Axis in Stallions." *Journal of Reproductive Fertility* Suppl. 27 (1979):67–73.

Ganjam, V. K., and R. M. Kenney. "Androgens and Oenstrogens in Normal and Cryptorchid Stallions." *Journal of Reproductive Fertility* Suppl. 23 (1975):67.

Gebauer, M. R., B. W. Pickett, and E. E. Swierstra. "Reproductive Physiology of the Stallion, II: Daily Production and Output of Sperm." *Journal of Animal Science* 39 (1974):732.

———."Reproductive Physiology of the Stallion, III: Extra-gonadal Transit Time and Sperm Reserves." *Journal of Animal Science* 39 (1974):737.

Gebauer, M. R., B. W. Pickett, J. L. Voss, and E. E. Swierstra. "Reproductive Physiology of the Stallion: Daily Sperm Output and Testicular Measurements." *Journal of the American Veterinary Medical Association* 165 (1974):711–13.

Hoagland, T. A., K. M. Ott, J. E. Dinger, K. Mannen, C. O. Woody, J. W. Riesen, and W. Daniels. "Effects of Unilateral Castration on Morphologic Characteristics of the Testis in One-, Two-, and Three-Year-Old Stallions." *Theriogenology* 26 (1986):397–405.

Hughes, J. P., and R. G. Loy. "Artificial Insemination in the Equine. A Comparison of Natural Breeding and Artificial Insemination of Mares Using Semen From Six Stallions." *The Cornell Veterinarian* 60 (1970):463.

Jasko, D. J. "Evaluation of Stallion Semen." *Veterinary Clinics of North America—Equine Practice* 8 (1991):129–48.

Jasko, D. J., H. R. Sawyer, and E. L. Squires. "Identification of Degenerative Germ Cells in Semen From a Quarter Horse Stallion." *Journal of Equine Veterinary Science* 5 (1991):283–86.

Johnson, L. "Increased Daily Sperm Production in the Breeding Season of Stallions Is Explained by an Elevated Population of Spermatogonia." *Biology of Reproduction* 32 (1985):1,181–90.

Johnson, L., and W. B. Neaves. "Age Related Changes in the Leydig Cell Population, Seminiferous Tubules and Sperm Production in Stallions." *Biology of Reproduction* 24 (1981):703–12.

Johnson, L., and D. L. Thompson. "Age-Related and Seasonal Variation in the Sertoli Cell Population, Daily Sperm Production and Serum Concentrations of Follicle-Stimulating Hormone, Luteinizing Hormone and Testosterone in Stallions." *Biology of Reproduction* 29 (1983):777–79.

Kenney, R. M., R. S. Kingston, A. H. Rajamannon, and C. F. Ramberg. "Stallion Semen Characteristics for Predicting Fertility." *Proceedings American Association of Equine Practitioners* 17 (1971):53.

Kenney, R. M., and W. L. Cooper. "Therapeutic Use of a Phantom for Semen Collection From a Stallion." *Journal of the American Veterinary Medical Association* 165 (1974):706.

Kirkpatrick, J. F., L. Wiesner, R. M. Kenney, V. K. Ganjam, and J. W. Terner. "Seasonal Variation in Plasma Androgens and Testosterone in the North American Wild Horse." *Journal of Endocrinology* 72 (1977):237.

Kosiniak, K. "Characteristics of the Successive Jets of Ejaculated Semen of Stallions." *Journal of Reproductive Fertility* Suppl. 23 (1975):59.

Long, P. L., et al. "Identification of Immature Germ Cells in Semen of Stallions." *Equine Practice* 15, no. 5 (1993):29–33.

Love, C. C., M. C. Garcia, F. R. Riera, and R. M. Kenney. "Evaluation of Measures Taken by Ultrasonography and Calipers to Estimate Testicular Volume and Predict Daily Sperm Output in Stallions." *Journal of Reproductive Fertility* Suppl. 44 (1991): 99–105.

Martin, J. C., E. Klug, and A. R. Gunzel. "Centrifugation of Stallion Semen and Its Storage in Large-Volume Straws." *Journal of Reproductive Fertility* Suppl. 27 (1979):49.

McCall, J. P., and A. M. Sorensen Jr. "Physiology of Properties of Stallion Semen." Abstract in *Journal of Animal Science* 30 (1970):329.

———."Evaporated Milk As an Extender for Stallion Semen." *A.I. Digest* 29, no. 12 (1971):8–10.

Merkt, H., K. O. Jacobs, E. Klug, and E. Aukes. "An Analysis of Stallion Fertility Rates (Foals Born Alive) From the Breeding Documents of the Landgestut Celle Over a 158-Year Period." *Journal of Reproductive Fertility* Suppl. 27 (1979):73.

Nash, J. G. Jr., J. L. Voss, and E. L. Squires. "Urination During Ejaculation in a Stallion." *Journal of the American Veterinary Medical Association* 176 (1980):224.

Nishikawa, Y. *Studies on Reproduction in Horses.* Tokyo: Japan Racing Association, Shiba Tamuracho Minatoku, 1959.

Nishikawa, Y., and S. Shinomeija. "Our Experimental Results and Methods of Deep Freezing of Horse Spermatozoa." *Animal Reproduction and AI* (1972):207–13.

Noiles, E. E., T. A. Hoagland, and J. E. Dinger. "Effect of Exercise and Ejaculation on FSH, LH, and Testosterone Concentrations in the Two-Year-Old Stallions." *Journal of Animal Science* 59, Suppl. 1 (1984):356.

Nyman, M. A., J. Geiger, and J. W. Goldzieher. "Biosynthesis of Estrogen by the Perfused Stallion Testis." *Journal of Biological Chemistry* 234 (1959):16.

Pickett, B. W. "Evaluation of Stallion Semen." In *Lameness in Horses*, edited by O. R. Adams, 60. 3d ed. Philadelphia: Lea and Febiger, 1973.

———."Stallion Management With Special Reference to Semen Collection, Evaluation and Artificial Insemination." In *Proceedings of the First National Horsemen's Seminar*, 37–47. Fredericksburg: Virginia Horse Council, 1976.

Pickett, B. W., et al. "Management of the Stallion for Maximum Reproductive Efficiency, II." *Colorado State University Animal Reproduction Laboratory Bulletin* No. 5 (1989).

————."Total Scrotal Width (TSW) of Normal and Abnormal Stallions." In *Proceedings of 33rd Annual Convention of the American Association of Equine Practitioners*, 1987.

Pickett, B. W., and D. G. Back. "Procedures for Preparation, Collection, Evaluation, and Insemination of Stallion Semen." *Colorado State University Experiment Station and Animal Reproduction Laboratory General Series* 935 (1973).

Pickett, B. W., L. C. Faulkner, G. E. Seidel Jr., W. E. Berndtson, and J. L. Voss. "Reproductive Physiology of the Stallion, VI: Seminal and Behavioral Characteristics." *Journal of Animal Science* 43 (1976):617–25.

Pickett, B. W., L. C. Faulkner, and T. M. Sutherland. "Effect of Month and Stallion and on Seminal Characteristics and Sexual Behavior." *Journal of Animal Science* 31 (1970):713–28.

Pickett, B. W., J. J. Sullivan, and G. E. Seidel Jr. "Reproductive Physiology of the Stallion, V: Effects of Frequency of Ejaculation on Seminal Characteristics and Spermatozoal Output." *Journal of Animal Science* 40 (1975):917–23.

Pickett, B. W., J. L. Voss, and L. K. Nelson. "Factors Influencing the Fertility of Stalion Spermatozoa in an A.I. Program." In *Proceedings of the 8th International Congress on Animal Reproduction*, A.I. IV: 1049, 1976.

Pickett, B. W., J. L. Voss, and E. L. Squires. "Impotence and Abnormal Sexual Behavior in the Stallion." *Theriogenology* 8 (1977):329.

Rasbech, N. O. "Ejaculatory Disorders of the Stallion." *Journal of Reproductive Fertility,* Suppl. 23 (1975):123.

Richardson, G. F., and M. S. Wenkoff. "Semen Collection From a Stallion Using a Dummy Mount." *Canadian Veterinary Journal* 17 (1976):177.

Skinner, J. D., and J. Bowen. "Puberty in the Welsh Stallion." *Journal of Reproductive Fertility* 16 (1968):133.

Squires, E. L., B. W. Pickett, and R. P. Amann. "Effect of Successive Ejaculation on Stallion Seminal Characteristics." *Journal of Reproductive Fertility* Suppl. 27 (1979):7–12.

Squires, E. L., G. E. Todter, W. E. Berndtson, and B. W. Pickett. "Effect of Anabolic Steroids on Reproductive Function of Young Stallions." *Journal of Animal Science* 54 (1982):576–82.

Swerczek, T. W. "Immature Germ Cells in Semen of Thoroughbred Stallions." *Journal of Reproductive Fertility* Suppl. 23 (1975): 135–37.

Swierstra, E. E., M. R. Gebauer, and B. W. Pickett. "The Relationship Between Daily Sperm Production as Determined by Quantitative

Testicular Histology and Daily Sperm Output in the Stallion." *Journal of Reproductive Fertility* Suppl. 23 (1975):35–41.

————."Reproductive Physiology of the Stallion, I: Spermatogenesis and Testis Composition." *Journal of Reproductive Fertility* 40 (1974):113–23.

Thompson, D. L. Jr., B. W. Pickett, W. E. Berndtson, J. L. Voss, and T. M. Nett. "Reproductive Physiology of the Stallion, VIII: Artificial Photoperiod, Collection Interval and Seminal Characteristics, Sexual Behavior and Concentrations of LH and Testosterone in Serum." *Journal of Animal Science* 44 (1977):656–64.

Thompson, D. L. Jr., B. W. Pickett, and T. M. Nett. "Effect of Season and Artificial Photoperiod on Levels of Estradiol-17b and Estrone in the Blood Serum of Stallions." *Journal of Animal Science* 47 (1978):184.

Thompson, D. L. Jr., B. W. Pickett, E. L. Squires, and R. P. Amann. "Testicular Measurements and Reproductive Characteristics in Stallions." *Journal of Reproductive Fertility* Suppl. 27 (1979):13.

Tischner, M., K. Kosiniak, and W. Bielanski. "Analysis of the Pattern of Ejaculation in Stallions." *Journal of Reproductive Fertility* 41 (1974):329.

Wierzbowski, S. "Ejaculatory Reflexes in Stallions Following Natural Stimulation and the Use of the Artificial Vagina." *Animal Breeding Abstracts* 26 (1958):367.

Wierzbowski, S., and E. S. E. Hafez. "Analysis of Copulatory Reflexes in the Stallion." *Proceedings of the 4th International Congress on Animal Reproduction.* A. I. 2 (1961):176.

INDEX

A

ampulla, 99
appraising, 167, 186

B

barns, 137
beans, 105
biting, 7, 23, 50, 213
bolting, 25
book of mares, 61, 130, 132, 156
breeding halter, 19
  (also see war bridle)
breeding incentives, 156
breeding roll, 104
breeding shed, 141

bulbourethral glands, 102
bulbourethral muscle, 102

C

Cells of Leydig, 95
  (also see testosterone)
charging, 25
chorionic gonadotropin, 131
clumping, 120
coagulating glands, 100
cod liver oil, 151
cold shock, 83
colliculus seminalis, 99, 102
concentration, 115
condoms, 66
conformation, 164, 181